Textile Science and Clothing Technology

Series editor

Subramanian Senthilkannan Muthu, SGS Hong Kong Limited, Hong Kong, Hong Kong

More information about this series at http://www.springer.com/series/13111

Subramanian Senthilkannan Muthu
Editor

Textiles and Clothing Sustainability

Implications in Textiles and Fashion

 Springer

Editor
Subramanian Senthilkannan Muthu
SGS Hong Kong Limited
Hong Kong
Hong Kong

ISSN 2197-9863   ISSN 2197-9871   (electronic)
Textile Science and Clothing Technology
ISBN 978-981-10-9552-8   ISBN 978-981-10-2182-4   (eBook)
DOI 10.1007/978-981-10-2182-4

Printed on acid-free paper

This Springer imprint is published by Springer Nature
The registered company is Springer Science+Business Media Singapore Pte Ltd.

Contents

Will Clothing Be Sustainable? Clarifying Sustainable Fashion

Sandra Roos, Gustav Sandin, Bahareh Zamani, Greg Peters and Magdalena Svanström

Abstract The Mistra Future Fashion research programme (2011–2019) is a large Swedish investment aimed at reducing the environmental impact of clothing consumption. Midway into the programme, research results and insights were reviewed with the intent to see what picture appears from this interdisciplinary consortium, developed to address the multiple sustainability challenges in clothing consumption and the tools for intervention. Such tools comprise product design, consumer behaviour changes, policy development, business models, technical development, recycling, life cycle assessment (LCA) and social life cycle assessment (SLCA). This chapter quantifies the extent of the sustainability challenge for the apparel sector, via an analysis of five garment archetypes. It also considers to what extent different interventions for impact reduction can contribute in society's endeavour towards sustainability, in terms of staying within an "environmentally safe and socially just operating space", inspired by the planetary boundaries approach. In particular, the results show whether commonly proposed interventions are sufficient or not in relation to the impact reduction necessary according to the planetary boundaries. Also, the results clarify which sustainability aspects that the clothing industry are likely to manage sufficiently if the proposed interventions are realised and which sustainability aspects that will require more radical interventions in order to reach the targets.

Keywords Fashion · Textiles · Sustainability · Life cycle assessment (LCA) · Social life cycle assessment (SLCA) · Planetary boundaries

---

S. Roos (✉) · B. Zamani · G. Peters · M. Svanström
Chalmers University of Technology, 41296 Gothenburg, Sweden
e-mail: sandra.roos@swerea.se

S. Roos
Swerea IVF, Box 104, 431 22 Mölndal, Sweden

G. Sandin
SP Technical Research Institute of Sweden, Eklandagatan 86,
412 61 Gothenburg, Sweden

© Springer Science+Business Media Singapore 2017
S.S. Muthu (ed.), *Textiles and Clothing Sustainability*, Textile Science and Clothing Technology, DOI 10.1007/978-981-10-2182-4_1

1 Introduction

In this section, we attempt to describe the sustainability challenges faced by the fashion industry, their scale and the potential to surmount them. We adopt a life cycle thinking approach built on basic product life cycle assessment (LCA). This first section qualitatively describes the challenges and management tools for the fashion industry. The second section describes a LCA in which we scale up the impacts identified using product LCA to characterise the impacts of Swedish, American and Chinese fashion consumption. We then attempt to define sustainability in practical and quantitative goals for the fashion industry. In the fourth section of this chapter, we go on to evaluate the extent to which different interventions can enable industry to reach these goals, before concluding with a general discussion.

1.1 Systemic Challenges for Sustainable Fashion

According to the classic Brundtland definition of sustainability, sustainable development "meets the needs of the present without compromising the ability of future generations to meet their own needs" (World Commission on Environment and Development 1987). In this context, the whole idea of a sustainable fashion industry may seem paradoxical. How can an industry focused on "wants" manage to deliver human "needs" in the long term? While fundamentally we need *clothing* for protection from variations in the weather (as distinct from *fashion*), it is possible to think of a human need for personal communication and that this need has been expressed via our clothing choices over the recorded history of humankind, long before current environmental concerns about the negative impacts of fashion emerged. Issues which challenge the clothing industry today, such as the excess use of limited resources, the pollutants the industry releases and the working conditions within it, are consequences of the modern scale and methods of the industry, so sustainable fashion ought to be possible, if we can decouple it from its current issues.

Limited resources are a key challenge to the sustainability of the clothing industry. Cotton is a very water-intensive product, with some analysts estimating that a kilogram of cotton textile demands the use of 8.5 tonnes of water (Pfister et al. 2009). Consequently, the production of cotton is limited by the availability of irrigation water in producer countries, where it must compete with food and fodder crops. To cope with this limited resource, the key alternative filling the demand for fibre is polyester derived from fossil hydrocarbons (Peters et al. 2014). "Peak oil" may seem like a distant prospect given the rise of the American fracking industry and current depressed oil prices (around USD 40 per barrel), but this leads to another problem, that of pollution. Our increasing reliance on synthetic fibres raises concerns associated with greenhouse gas emissions in the supply chain. Where waste textiles are combusted, this issue is also present, elsewhere the problem may be limited landfill space or the micro-plastic pollution caused (in part) by the emission of textile fibres to the environment (Eerkes-Medrano et al. 2015). There is also a wide range of other chemical emissions associated with textiles, of which the durable water-proofing chemicals are perhaps the most persistent (Holmquist et al. 2016).

Working conditions are another challenge for the sustainability of the clothing industry. The globalisation of the clothing industry over the last 30 years has perhaps lifted many people in Asia from poverty, but it has also created concerns about the abuse of labour in countries that do not have strong labour representation. This ranges from overwork, gender discrimination, child labour and (most notoriously in the case of the Rana Plaza disaster in Bangladesh) unsafe working conditions. In such cases, the sustainability problem is not connected with the ability of future generations to meet their needs, but with the current generation. A systemic aspect of the problem is that the supply chains are now so globalised and complex that it can be difficult for managers, for example those in a European clothing retailer, to know precisely where the garments they sell are being produced. The long supply chain between buyer, contractor and many subcontractors isolates management from distant workplaces, so that even if consumers demand better conditions for garment production, their implementation is challenging.

1.2   The Idea of Sustainability Assessment

There are several analytical difficulties associated with the desire to create a sustainable clothing industry. As suggested in the previous section, long supply chains are one of them. The idea of life cycle thinking is intended to deal with this difficulty, by ensuring that analysts consider the entire life cycle of a product, from cradle to grave. Environmental LCA is a tool based on this thinking, and the ISO standard for this method (ISO14040) (ISO 2006) provides a fundamental basis for attempts to apply it to fashion products and the clothing industry.

LCA is a globally used and accepted method for assessing environmental impacts of a product's life cycle from cradle to grave, including raw material extraction, material processing, product manufacture, distribution, use, disposal and recycling (European Commission 2014). This extended conception of the technical system under study means that it is in principle able to avoid suboptimisation. In other words, an LCA should be able to tell you if a garment is achieving better environmental performance by minimising impacts during production, but at the expense of greater impacts at its end of life.

LCA has another particular characteristic that aids its use in sustainable development—it has a broad conception of the environment. While a water footprint may be a useful proxy for several kinds of impacts on human health, resources and the environment, LCA can employ a larger suite of indicators reflecting a technical system's impacts via climate change, stratospheric ozone layer depletion, tropospheric ozone pollution, land use change, energy consumption, toxic emissions and more. This breadth in the conception of "the environment" in LCA means that the risk of "problem shifting" is reduced. In other words, if you have carbon compensated a garment but that means greater water consumption, an LCA should tell you this.

Social LCA is an attempt to ensure that (for example) working conditions in the clothing industry can be evaluated in a framework consistent with life cycle thinking (Muthu 2015a, b). As a scientific field, it is relatively recent compared with environmental LCA. A recent review of SLCA selected 2006 as a starting point (Garrido et al. 2015) and the UNEP/SETAC guidelines for SLCA were only published in 2009 (UNEP 2009). On the other hand, the original ISO standard for environmental LCA was first published in 1997. SLCA has been used in several industries, but at the time of writing there were no published peer-reviewed SLCAs of the clothing industry.

Ecological footprint analysis is another approach to environmental sustainability assessment. An ecological footprint is defined as the area of land necessary to support a population indefinitely (Rees and Wackernagel 1996; Wackernagel et al. 1999). In brief, this area includes the land necessary to produce the food, fibre and other resources humans consume, plus the land necessary to soak up our carbon pollution. Ecological footprints have certain advantages over LCAs, the key one being ease of communication. It is relatively easy to explain to laypersons that their lifestyle requires a certain number of (simply visualised) *hectares*, compared with the challenge of to a number of (invisible) *kilograms of carbon dioxide equivalent greenhouse gas*, which is just one possible LCA indicator. Furthermore, by comparing an ecological footprint with the area of land available on the planet, an absolute comparison can be made. By making this comparison, the Global Footprint Network has drawn policymakers' attention to necessary changes we need to make towards sustainability (GFN 2016).

Of course ecological footprinting is open to criticism. A key aspect that concerns LCA analysts is that in mainstream ecological footprints, only the impacts of carbon pollution are quantified. For many engineered systems, a wide range of other contaminants are thought to pose significant threats to human health and the environment (Peters et al. 2008). One way to deal with this would be to broaden ecological footprinting. Another would be to attempt to make comparisons between LCA indicators and global sustainability thresholds. The original focus of the studies that came to be called LCA was at the product scale (particularly food packaging) (Baumann and Tillman 2004), which of course means evaluation of product supply chains. Later in this chapter, we examine the potential to scale up the results of product LCA and SLCA to the scale of a whole industry sector and compare the results of this assessment with global sustainability goals.

1.3 Cyclonomy or Bust? Possible Ways Ahead

How can the clothing industry change towards sustainability? There are external forces driving minimum standards for certain issues. For example, European legislative developments on the management of the risks associated with persistent chemicals (e.g. the REACH legislation) affect the clothing industry among others. There are a number of initiatives within the industry that are striving towards that goal. Predictably in an industry with such long supply chains, the use of certification schemes is growing fast in the clothing industry. There is a strong focus on business-to-business disclosure schemes such as the one developed by Zero Discharge of Hazardous Chemicals, an industrial collaboration (ZDHC 2014). There are many ecolabelling schemes that are relevant to the clothing industry (Clancy et al. 2015), including the EU Ecolabel, Bluesign, Cradle to Cradle, Global Organic Textile Standard and the OEKO-TEX® Standard 100 and Made in Green. One of the problems facing consumers is the wide variety of labels, which can lead to confusion. Another is that certain labels only take particular issues into account. There are attempts to create more comprehensive labelling schemes, for example as part of the Higg Index tool under development by the Sustainable Apparel Coalition (SAC 2016). At the moment, this is a set of modules for social and environmental sustainability self-assessment of labels, facilities and products, but the intent of the consensus process behind the Higg Index is to also produce a consumer-facing sustainability label for products. Another key development at this time is the growth of environmental product declarations (EPDs). These are typically used for business-to-business communication on account of the greater level of detail they communicate—an EPD is typically 15–25 pages in length. The rate of EPD registration has accelerated recently, and the European Union has driven the development of Product Environmental Footprint Category Rules. Developments such as these promise the availability of more useful information for corporate buyers and green public procurement.

Inside and outside the industry, there are plenty of voices calling for more radical rethinking of the business model. Achieving more efficient use of resources means reducing demand, reusing resources and recycling. In an industry driving towards ever faster fashion, these can seem to be radical ideas (Carbonaro and Goldsmith 2015). But the idea that the *average* Swede buys 9 T-shirts per year (Roos et al. 2015b) as a consequence of fast fashion would probably be regarded as radically wasteful by earlier generations. Generally speaking, in industrialised countries, we consume more clothes at a higher rate than their technical lifespans demand. So buying less is a possible response, although it runs the potential risk of rebound effects (Font Vivanco et al. 2015). An interesting manifestation of this response to waste was Patagonia's "Don't buy this jacket" campaign, challenging consumers to stop spending (Carbonaro and Goldsmith 2015). Another response is to try and slow down fashion consumption by spending more money per garment on fewer, higher-quality garments and repairing them when appropriate. This could represent a return to mid-twentieth-century norms for Western consumers. A possible alternative for the consumer addicted to rapidly changing fashion is the development of collaborative consumption models for clothing (Zamani et al. 2016a). An example of this is the development of clothing libraries where the idea is that for a monthly membership fee, the member may borrow a certain number of garments. Rental businesses for formal attire have existed for decades, but applying this form of garment ownership to the rest of the wardrobe is innovative. Recycling is a general term which may extend from the well-established garment reuse business models, such as those employed by charity organisations to provide garments to consumers in developed and developing countries, to physical shredding and reconstruction of textiles (Recover 2016), to novel chemical dissolution and recycling pathways that have yet to be established at industrial scale (Roos et al. 2015b). These paths to a more circular economy via recycling retain traditional notions of garment ownership that are abandoned in the collaborative consumption model.

1.4 Multidisciplinary Problem Solving

The need for more multidisciplinary research on pathways to sustainability for the fashion industry became increasingly apparent to Swedish research managers in the first years of the new millennium. In response to this need, the Swedish Foundation for Strategic Environmental Research ("Mistra") created a directed research programme known as "Mistra Future Fashion". Thus, a collaboration between several disciplines including textile engineering, consumer psychology, political science, textile design, business management and environmental systems analysis was created. Collaboration and dialogue with industrial actors such as retailers, public procurers and recyclers were part of the programme. Midway into the programme (2015), research results and insights were reviewed with the intent to see what picture appears from this interdisciplinary consortium, developed to address the multiple sustainability challenges in clothing consumption and the tools for intervention (Mistra Future Fashion c/o SP 2015).

2 Environmental and Social Impacts, from Garment Level to Industry Sector Level

This section describes the challenge of environmental and social life cycle assessment (LCA and SLCA) in the textile sector, the availability and relevance of methods for this, the results from a case study on the Swedish clothing industry, and the generality and representativeness of such a study.

In LCA, as well as in many other scientific areas, the design of the study decides the type of answers that can be retrieved from a study (Baumann and Tillman 2004). The choice of system boundary must thus be selected with great care for each study. Sandén and Hedenus (2014) describes, for example, the opportunities and challenges with making assessments of bio-based fuels with different system boundaries applied, from very narrow to very wide. The assessment can be narrow and limited to, e.g., ethanol production. Or, the assessment can take a wider approach, looking at, in this case, fuel production (enabling comparison with other fuels), or an even wider approach looking at vehicle propulsion (enabling comparison with other propulsion technologies), and so forth, all the way up to the assessment of environmental impacts of final end uses, e.g. passenger transport, or even communication. Depending on the purpose of the study, all these levels may be applicable. However, for strategic decisions that are typically more long term, society wide and strategic, a wider system boundary will be more relevant.

Too inclusive system boundaries may also be a problem. In LCA, there is always a trade-off between completeness in scope and completeness in data representativeness. Completeness in scope comes at the price of simplifications and uncertainties, stated recently Hellweg and Milà i Canals (2014). When an assessment is made of a technological process in general instead of a specific process at a specific plant, the issue of representativeness is crucial. Sandén (2012) suggests that a way to handle this is to be explicit with the division into the foreground system, i.e. the technology in focus of the study, and the background systems, i.e. the context (state, situation) the foreground system is placed in. This section will elaborate on the dependence on system boundaries in the environmental evaluation of textile products and the representativeness of a case study for drawing general conclusions.

The purposes of our recent LCA and SLCA studies (which will be used as a case study throughout this chapter) were to deliver knowledge and solutions that the Swedish clothing industry and its stakeholders can use to significantly improve the clothing sector's sustainability performance and strengthen its global competitiveness (Zamani et al. 2016b). The sector perspective was thus central for the goal and scope of the studies (Roos et al. 2015b). To approach the challenge of environmental and social assessments of a whole sector, while still providing credible data for technological processes on a general level, an industry sector approach was developed (Roos et al. 2016). This section summarises key results from the LCA and SLCA studies and the industry sector approach.

2.1 Life Cycle Assessment, from Garment Level to Industry Sector Level Environmental Performance

The analysis of the environmental impact of an entire industrial sector with LCA would be an immense task, and some simplifications were made in the study of the Swedish textile sector (Roos et al. 2015b). Five garment archetypes were selected for the study with the primary aim that they would be representative in terms of fibre types and textile production methods for Swedish clothing consumption and also public sector procurement. Additionally, the intent was to choose garments with sufficiently diversified life cycles so that they would be able to show the significance of interventions in different life cycle phases for different types of garments. For example, T-shirts are washed more often than jackets. T-shirts are therefore expected to exemplify the value of changes within the use phase, while jackets are expected to more clearly exemplify the value of changes to the garment life cycle outside the use phase.

2.1.1 Selected Garments and Their Life Cycle Models

Four fashion garments and one garment for the public sector were selected: a T-shirt, a pair of jeans, a dress, a jacket, and a hospital uniform. Each of these garments is a common high-volume product that consists of materials that are used also for other types of garments and can thus represent also other garments (see Table 1).

In the study context, another foreseen purpose with the results was to be able to answer the question about the potential consequences of commonly proposed interventions for impact reduction (see Sect. 4), and therefore, the process technology was chosen to be quite modern. This means that Best Available Technology (BAT) or close-to-BAT processes were modelled for textile processes. If we instead would have chosen to model technology that is known to be outdated and will be replaced within the next few years, the answer to what interventions are needed would be to change to BAT and that would have reduced the value of the study. This choice means that the environmental impact of the Swedish clothing consumption was slightly underestimated.

By limiting the study to five garments, it was possible to model the life cycles of these garments in a sufficient level of detail for the study, keeping the resolution high enough to, for example, be able to see the results from interventions.

Table 2 shows an overview of the processes that were included as a result of the selection of the five garments. The processes may also be varied depending on garment; for example, wet treatment, dyeing and printing will naturally be different for all five garments. For more details of each process, see Roos et al. (2015b).

Table 1 Properties of the five garment archetypes selected for the study of the Swedish textile sector

	T-shirt	Jeans	Dress	Jacket	Uniform
Mass	110 g	477 g	478 g	444 g	340 g
Textile material	100 % cotton	98 % cotton 2 % elastane	100 % polyester	44 % polyamide 48 % polyester 18 % cotton/elastane mix	50 % cotton 50 % polyester
Construction	Knitted	Woven	Knitted/woven	Knitted/woven	Woven
Number of uses	22	200	10	100	75
Use phase	Washed after 2 uses. 34 % dried with heat 15 % ironed	Washed after 10 uses. 29 % dried with heat 41 % ironed	Washed after 3 uses. 19 % dried with heat 18 % ironed	Washed once. 21 % dried with heat 5 % ironed	Washed after 1 use. 100 % dried with heat 0 % ironed
End of life	Municipal incineration with cogeneration of heat and electricity				

Table 2 Processes included in the study. The "X" shows for which garments the process is included in the life cycle

Process	T-shirt	Jeans	Dress	Jacket	Hospital uniform
Production phase					
Cotton cultivation, ginning and baling	X	X	–	X	X
Polyester fibre production	–	–	X	X	X
Polyamide fibre production	–	–	–	X	–
Elastane fibre production	–	X	–	X	–
Yarn spinning	X	X	X	X	X
Knitting	X	–	X	X	–
Weaving	–	X	X	X	X
Non-woven process	–	–	–	X	–
Wet treatment, dyeing and printing	X	X[a]	X	X	X
Product assembly	X	X	X	X	X
Distribution and retailing phase					
Distribution and retail	X	X	X	X	–
Industrial laundry distribution	–	–	–	–	X
Use phase					
User transportation, to and from the store	X	X	X	X	–
Residential washing	X	X	X	X	–
Residential drying	X	X	X	X	–
Residential ironing	X	X	X	X	–
Use of T-shirt (22 uses)	X	–	–	–	–
Use of jeans (200 uses)	–	X	–	–	–
Use of dress (10 uses)	–	–	X	–	–
Use of jacket (100 uses)	–	–	–	X	–
Use of hospital uniform (75 uses)	–	–	–	–	X
Industrial laundry	–	–	–	–	X
End of life phase					
Incineration	X	X	X	X	X

[a]Please note that the wet treatment for jeans material is made on yarn and is performed prior to the weaving

2.1.2 Scaling up to Swedish National Clothing Consumption

The upscaling was performed using statistics on Swedish import, export and domestic production for 2012 (Statistics Sweden 2014). Statistics Sweden divides products after the Combined Nomenclature (CN) used by the customs. Textile garments are divided into tricot textile garments (CN code 61) and non-tricot textile garments (CN code 62) and then follow a division into 34 different garments' classification groups (fibre type and garment type, men/women's wear, etc.) that each can be subdivided into subcategories. The groups that each garment represents were selected based on the following prioritisation of criteria:

1. Knitted or woven construction,
2. Fibre type (cotton, synthetics, regenerated or denim),
3. Similarity, in terms of function of the garment, use pattern, etc.

Table 3 shows for each modelled garment how large part of the Swedish clothing consumption was assumed to be represented by each garment.

The industry sector approach enabled national statistics on imports, exports and production to be used as a basis for the assumptions about the length of the garment service lives. Thus, the garment service lives were not based on hypothetical technical service lives, but on the actual user behaviour as indicated by national statistics [combined with some complementary assumptions, see Roos et al. (2015b) for more details]. This means that the assumed number of uses per garment reflects average behaviour by Swedes. It should be noted that the number of uses in this study does not correspond to the number of washing cycles, as the number of washing cycles per use was also based on data on actual consumer behaviour, collected by means of consumer surveys (Granello et al. 2015; Gwozdz et al. 2013).

2.1.3 Results

This section shows selected results, focusing on climate impact calculated with the GWP100 method (IPCC 2013). For more results, see Roos et al. (2015b). The study of the environmental impact *per use* of each of the five garments (the functional units of the study) allows for studying micro-level aspects, such as the environmental significance of different life cycle phases for different garments. The assessment of the environmental impact of *the total Swedish clothing sector* is a study of macro-level aspects, such as the relative importance of different garments and the potential of different measures for impact reduction.

Figure 1 below shows the potential climate impact *per use*. The polyester dress has a relatively high impact per use, which is because a dress is, according to the statistics, used much fewer times per service life compared to the other garments.

Table 3 The representation of Swedish clothing consumption based on 2012 statistics

Garment	Volume (tonnes)	Percentage of the modelled Swedish clothing consumption (%)
T-shirt	19,672	24
Jeans	16,138	19
Dress	21,518	26
Jacket	20,500	25
Hospital uniform	5604	7
Total	83,432	100

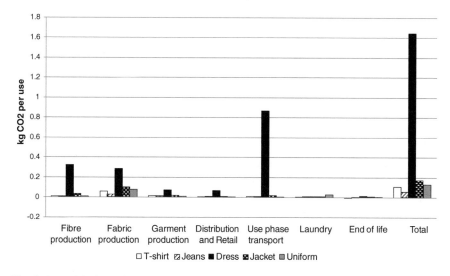

Fig. 1 Potential climate impact in kg CO_2 equivalents per statistical use of garment

Figure 1 also shows that the laundry phase is close to insignificant for all garments, a result that is further discussed in Sect. 2.3.

Figure 2 shows the impact per garment service life, i.e. not divided with the number of uses. Here, the polyester dress comes out much better, and it is instead the jacket that bears the highest environmental burden.

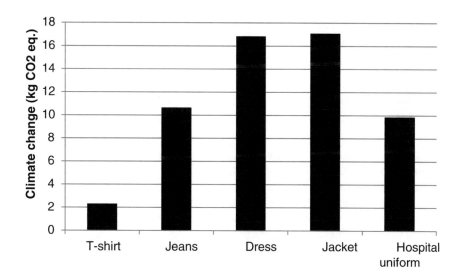

Fig. 2 Potential climate impact in kg CO_2 equivalents per service life of garment

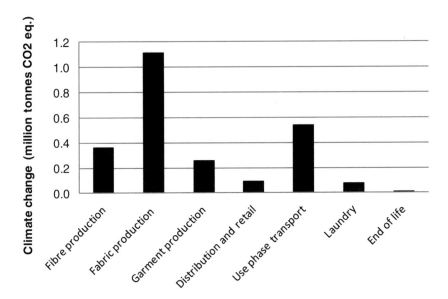

Fig. 3 Potential climate impact in million tonnes CO_2 equivalents for total clothing purchases in Sweden over one year

Finally, the result on national level is shown in Fig. 3. The total climate change contribution from the clothing industry sector in Sweden is around 2.3 million tonnes carbon dioxide equivalents per year. The Swedish population in 2012 was 9,555,893 people (Statistics Sweden 2014), which means that the carbon footprint from clothing consumption is around 0.25 tonnes CO_2 equivalents per capita and year. The average carbon footprint for a Swede is around 10 tonnes of CO_2 equivalents per year (Larsson 2015), which means that the carbon footprint share from clothing is around 2.5 % today.

2.2 Social Impact Assessment of the Swedish Clothing Industry Sector

In this study, a set of social indicators was first selected based on a survey used to examine priorities of Swedish clothing consumers (Zamani 2014). For identifying the social hotspots for this set of indicators, a cradle-to-grave SLCA was carried out using the Social Hotspots Database (SHDB) (Benoit-Norris et al. 2012). The SHDB provides data on potential social risk levels (low, medium, high and very high) in different sectors at different geographical locations. To identify the product's supply chain, an input/output (I/O)-based database provided by the Global Trade Analysis Project was used. Further, the social impact of the purchase of 1 USD of garments was evaluated in terms of the number of work hours associated with a certain level of risk of negative social impacts for each sector in a specific country. High and very high risk levels of each social indicator were considered as social hotspots where social targets are not met, in line with (Ekener-Petersen et al. 2014). The results allow us to identify the negative social hotspots on industry sector level. In Zamani et al. (2016), we focused on social risks associated with Swedish clothing consumption. Here, we also add results for the USA and China, to show that the approach is applicable also in studies concerned with other geographical contexts.

2.2.1 Results

Figures 4, 5 and 6 show the resulting risk-level intensity (social target transgression intensity) for each indicator for yearly consumption in Sweden, USA and China. For more results for the Swedish context, see Zamani et al. (2016b).

Results from social hotspot identification in Figs. 4, 5 and 6 show the significant social risks in textile and clothing industry related to wage, child labour and safe working conditions and gender inequality. The risk-level intensity for Swedish and Chinese consumption showed to be highest for the risks of "wage under 2 USD" and "child labour". The highest social risks in textile and clothing supply chain for consumption in the USA are related to "fatal injury" and "child labour".

2.3 Use Phase Assumptions and National Statistics

Figure 2.1 shows that the laundry in the use phase is close to insignificant for the Swedish clothing sector. This is a contrary finding to several previous studies pointing out the use phase as dominating on climate impact (Allwood et al. 2006; BIO 2007; Levi Strauss & Co. 2015). LCA studies of garments therefore often

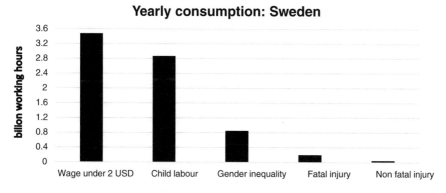

Fig. 4 Very high and high-risk-level intensity for each social indicator based on work hours for yearly consumption in Sweden

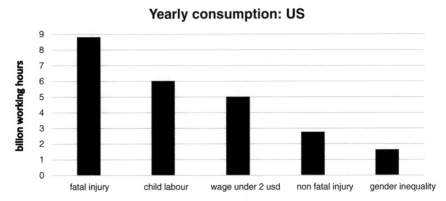

Fig. 5 Very high and high-risk-level intensity for each social indicator based on work hours for yearly consumption in the USA

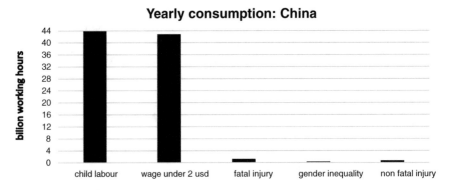

Fig. 6 Very high and high-risk-level intensity for each social indicator based on work hours for yearly consumption in China

investigate the climate saving in lowering the washing temperature from 60 to 40 °C (Beton et al. 2014; Laursen et al. 2007), often disregarding that 60 °C is seldom used and that some garments (underwear, bedlinen, etc.) can require a high washing temperature to get a clean result (Krozer et al. 2011).

Table 4 shows the use phase parameters of our study, based on statistics of real user behaviour in Sweden (Roos et al. 2015b) and, as a comparison, examples from the literature of assumptions on user behaviour in terms of garment life length, number of washes, washing temperature and drying method. It is not surprising that LCA studies of a specific garment often assume the expected technical performance of the garment (theoretical life length), as the commissioner behind the study perhaps wants to signal that their products are of good quality. However, the clear differences between real and theoretical use phase parameters evident in Table 4 suggest that such LCA studies do not reflect real garment life cycles. Assuming the

Table 4 Use phase parameters in LCA studies of garments

Study	Garment	Life length	Number of washes (base case assumption)	Washing temperature	Tumble drying/Drying cabinet
Use phase parameters based on statistics of real user behaviour					
Roos et al. (2015b)	T-shirt	22 uses	11 times	40 °C	34 % of washing cycles
Roos et al. (2015b)	Jeans	200 uses	20 times	40 °C	29 % of washing cycles
Roos et al. (2015b)	Dress	10 uses	3 times	40 °C	19 % of washing cycles
Roos et al. (2015b)	Jacket	100 uses	1 time	40 °C	21 % of washing cycles
Roos et al. (2015b)	Hospital uniform	75 uses	75 times	85 °C	100 % of washing cycles
Use phase parameters based on expected technical performance of the garment					
Levi Strauss & Co. (2015)	Jeans	2 years	104 times	warm[a]	100 % of washing cycles
Allwood et al. (2006)	T-shirt	n/a	25 times	60 °C	No
Bevilacqua et al. (2011)	Sweater	5 years	75 times	30 °C	no info
Laursen et al. (2007)	T-shirt	1 year	50 times	60 °C	100 % of washing cycles
Laursen et al. (2007)	Jogging suit	1 year	24 times	40 °C	100 % of washing cycles
Laursen et al. (2007)	Workwear jacket	3 years	40 times	85–90 °C	100 % of washing cycles
Laursen et al. (2007)	Viscose blouse	1 year	25 times	40 °C	No
BIO (2007)	Linen shirt	–	107 times	40 °C	No
BIO (2007)	Cotton shirt	–	107 times	40 °C	No
Collins and Aumônier (2002)	Cotton briefs	2 years	104 times	50 °C	50 % of washing cycles
Collins and Aumônier (2002)	Trousers	2 years	92 times	50 °C	50 % of washing cycles

[a]105 F (40 °C)

expected technical performance is particularly unsuitable when scaling up the LCA study from the garment level to the level of an entire industry sector, as it implicitly assumes that every garment is actually used its theoretical life length.

2.4 Generality and Representativeness for Other Contexts

The generality of the study of the Swedish clothing sector is limited due to several aspects, whereof the most important ones will be studied here. To increase the generality of the study and investigate whether more general conclusions can be drawn, the foreground system, i.e. the technology in focus of the study, has been placed in the context of two new countries: the USA and China, which gives scenarios with different background systems to the study. The aspects that are varied within the scenarios are as follows:

- The amount of garments consumed per capita,
- The electricity mix used for washing and drying garments in the use phase,
- The washing habits of the consumers,
- Consumer transports.

The production is assumed to have roughly the same environmental impact for all markets and is not varied.

2.4.1 Amounts of Garments Consumed Per Capita

The amount of garments consumed per capita, important for the life length of the garment, varies between countries. Table 5 shows the yearly national clothing consumption for Sweden (Statistics Sweden 2014), the USA and China (Wazir Advisors 2015).

For a translation into *kg of garment* consumed, it must be known how many kg of garments that one US dollar approximates in the different countries, and such statistics have not been found. Instead, a rough estimation of the generality was made, based on the Swedish statistics where both weight of garments (in kg) and price of garments (in SEK) are included, and 700 US dollars a year roughly corresponds to 10 kg a year or purchase of 50 new garments per year (Statistics Sweden 2014).

The purchase statistics for garments do not include second-hand purchase or home textiles such as towels and bedlinen.

Table 5 Clothing consumption in Sweden, China and the USA in 2013

	Total consumption (billion US dollar/year)	Consumption per capita (US dollar/year)
Sweden	6.7	704
USA	230	730
China	165	121

2.4.2 The Electricity Mix Used for Washing and Drying Garments in the Use Phase

Further, it is assumed that laundry uses national average electricity, and thus, the climate impact per kWh is 100, 530 and 970 g CO_2-eq. for Sweden, USA and China, respectively (Brander et al. 2011).

2.4.3 The Washing Habits of the Consumers

During the preparation work for the European Ecodesign Directive (2009/125/EC) (European Commission 2009), several preparatory studies were performed with the aim of comparing the climate change potential of different product groups, i.e. the aim was to achieve comparability between product groups. Washing machines (Stamminger 2005) and laundry dryers (Lefèvre 2009) were two of the product groups compared under the Ecodesign Directive development. The most commonly used washing temperature in the ten countries included in the study was 40 °C, and the average load was 3.2–3.3 kg laundry/cycle. The average energy consumption increase per degree Celsius increase was 0.03 kWh/°C.

Table 6 below reports the comparative energy consumption of washing machines and tumble dryers in the average scenarios for European countries.

In Table 6 is shown that the average energy consumption from tumble drying is 3–4 times higher than the energy consumption of washing at 40 °C. It is also clear that refraining from tumble drying saves more energy than lowering the washing temperature, though this has some impact.

In the study by Roos et al. (2015b), specific Swedish data for consumer laundry were retrieved from the studies of Gwozdz et al. (2013) and Faberi (2007), showing that 40 °C was the most common washing temperature and tumble drying was performed after roughly 20 % of the washing cycles. The washing temperature was set to 40 °C for all countries, while the frequency of tumble drying was

Table 6 Energy consumption of washing machines and tumble dryers	Energy consumption per kg laundry (kWh)
Wash 30 °C	0.15
Wash 40 °C	0.21
Wash 60 °C	0.39
Wash 90 °C	0.60
Air vented tumble dryer (class C)	0.67
Condenser tumble dryer (class C)	0.73

parameterised, and set to occur after on average 20, 100 and 0 % of the washes in Sweden, USA and China, respectively.

2.4.4 Consumer Transports

The consumer transport is assumed to be a mix of modes in Sweden [on average 3.85 kg CO_2-eq./kg garment (Roos et al. 2015b)], car in the USA (on average 5 kg CO_2-eq./kg garment) and by foot/bicycle in China (on average 0 kg CO_2-eq./kg garment), the second two figures chosen just to give clear results.

2.4.5 Summary of Scenario Parameters

Table 7 summarises the parameters used in the study of the Swedish clothing sector and for the scenario development for USA and China.

2.4.6 Results from the Scenarios

Table 8 shows how the percentage between the climate impact on production, consumer transport and laundry varies in the scenarios.

Scenario A represents the current states for Sweden and USA, where roughly 10 kg of garments per person and year is consumed (consumption per capita in US dollar/year is 704 respective 730 (see Table 5)). As discussed in 2.4.1, the consumption per capita in China of 121 US dollars/year is difficult to translate into kg of garments, and scenario B as well as C might apply.

The use phase laundry contribution to climate impact on the garment life cycle clearly increases as the climate impact on the national electricity mix increases. Also, the life length (or rather the rate of consumption of new garments) plays a significant role for the distribution between production and laundry climate impacts.

Table 7 Parameters in the study of the Swedish clothing sector and for the scenario development for USA and China for the use phase

Use phase parameters	Sweden	USA	China
Amounts of garments consumed per capita (USD/year)	704	730	121
Electricity climate impact (kg CO_2-eq./kWh)	40	530	970
Average washing temperature (°C)	40	40	40
Average washing frequency (uses per wash)	2	1	2
Tumble drying (drying cycles per wash)	0.2	1.0	0
Emissions for consumer transport (kg CO_2-eq./kg purchased garments)	3.85	5.0	0

Table 8 Scenarios for production, consumer transport and laundry in the different countries: Sweden, USA and China

	Production	Consumer transport	Laundry	Sum	Production percentage (%)	Laundry percentage
Scenario A: 10 kg a year						
Sweden*	70	27	3	100	70	3
USA*	70	35	40	145	48	27
China	70	0	73	143	49	51
Scenario B: 5 kg a year						
Sweden	35	14	3	52	68	6
USA	35	18	40	92	38	43
China*	35	0	73	108	32	68
Scenario C: 3 kg a year						
Sweden	24	9	3	36	67	8
USA	24	12	40	75	32	53
China*	24	0	73	97	25	75

*current states

3 Defining Sustainable Clothing from the Perspective of a Safe and just Operating Space

In steering towards a goal, it is important to understand and define that goal. In the case of the endeavour towards a more sustainable clothing industry, we must understand and define what we mean by a "sustainable clothing industry". Depending on the actor, and the actions that are within the actor's reach, this may, for example, be about understanding and defining a "sustainable garment", a "sustainable fashion company", a "sustainable national retail sector" or some other entity somewhere on the scale from the single garment to the global clothing sector. Sustainability must be understood and defined at the proper level.

Sustainable development is often defined as the simultaneous consideration of environmental protection, social well-being and equity and economic feasibility. In this section, we make an attempt to define sustainable clothing from the perspective of the clothing sector. As a starting point in understanding the environmental challenges, we use the popular and fairly recent descriptions and estimates of the planetary boundaries (Steffen et al. 2015). This framework was launched as a science-based (but naturally value-influenced) approach to understanding global limitations of the ecosystem functions and the extent of human-induced pressures. By translating these boundaries to the clothing product level, we believe that the goal in terms of the environmental impact of the future clothing sector becomes understandable. For social well-being and equity, it is much harder to define boundaries in terms of the functioning and well-being of societies and individuals. We therefore rely on a qualitative discussion of goals recently agreed on a global level—some of the sustainable development goals (SDGs). In terms of economic feasibility, we adhere to more recent movements in the sustainable development discussion and view it as a tool rather than a goal in itself and to the idea of strong sustainability as we assume that there are limits to how different social, ecological, and economic foundations of our well-being are substitutable. For sure, the clothing sector would not choose a path that would not be economically feasible in the long term as the clothing sector, just as any business sector, is made up of "organisms" that are driven primarily (but not only) by economic values. Here, however, we want to focus on the goals for the clothing sector in terms of a safe and just operating space (Raworth 2012), which will be further described in Sect. 3.2.

3.1 Sustaining Ecosystem Functioning

Sustainability is often understood as a state of the planet in which we are not further diminishing the opportunities of current and future generations to meet their needs (World Commission on Environment and Development 1987). Among others, this can be interpreted as a state in which environmental degradation occurs at a rate that is slower than the rate at which nature can replenish itself and sustain a capacity to provide the ecosystem services needed by humanity. In the end, the notion of a sustainable state of the economy (not to be confused with a static state—we assume one can have development and change while still maintaining ecosystem functioning) suggests that there are *absolute* boundaries to how much pressure human society can impose on nature. This insight has led to the development of frameworks such as the ecological footprint (GFN 2016) and the planetary boundaries (Steffen et al. 2015), which relate human pressures on nature to nature's capacity to sustain such pressures. As mentioned in Sect. 1, the ecological footprint framework focuses on the productivity of land (and sea), by relating the available area to the area needed to produce the renewable resources used by humans and to sequester greenhouse gases emitted by humans. The planetary boundaries framework instead suggests nine biophysical global boundaries which human pressures cannot transgress unless inflicting risks of nonlinear changes leading to collapse of important ecosystem functions. In other words, the planetary boundaries framework suggests a "safe operating space" for human society to develop within. So, it provides a definition for a sustainable state of the planet, at least in environmental terms and at least for pressures acting on a global scale.

The ecological footprint metric and the planetary boundaries framework represent attempts to use science to identify a desired absolute state of sustainability. Also, there are political agreements based on keeping human pressures on natural systems within certain limits. Globally, examples of such agreements are the two-degree goal in international work for reducing climate change (UNFCCC 2016) and the Montreal Protocol for phasing out chlorofluorocarbons in order to protect the stratospheric ozone layer (United Nations 2016). On a national level, examples of agreements include the 17 Swedish environmental objectives, which guide national attempts of protecting various aspects of nature (Swedish EPA 2015). The political agreements are often influenced by science, but they are the result of political processes; thus, they accommodate compromises between different interests and different notions of what is politically possible to accomplish. In other words, they are heavily influenced by value-based considerations.

It should be noted that also more scientific attempts to define sustainability are influenced by value-based considerations, although perhaps not to the same extent as the political agreements. For example, the planetary boundaries are limits, which, if transgressed, inflict risks which the scientists behind the framework deem to be unacceptable. What to consider as an *unacceptable risk* is a value-based decision. In this case, the scientists have chosen to adopt the precautionary principle, i.e. that in case of uncertainty, we are better safe than sorry. Still, the outcome is certainly more science-based than the outcome of (more) political processes. It is, however, important to remember that defining sustainability will always include value-based considerations.

3.2 Fulfilling Human Needs in an Equitable Way

Social thresholds are even more difficult to define than the environmental thresholds. Our conceptions of a good life vary, and our expectations in life vary around the world and over time. Inspired by the millennium development goals and the discussions that eventually led to the SDGs, Kate Raworth discussed what she calls the social foundations of human well-being (Raworth 2012). Raworth created a framework that describes how humanity has to prepare to live within a doughnut-shaped space that she refers to as an "environmentally safe and socially just operating space" (see Fig. 7).

The planetary boundaries concept is commonly expressed as a circle (Steffen et al. 2015), where the normalised radius decides the pressure on the earth's carrying capacity for the different planetary boundaries. The outside arrows in Fig. 7 represent how the need to stay within the planetary boundaries pushes the outer circle inwards—we need to stay "within" a certain radius figuratively speaking. The demands for food, water, energy and other elements that make up the social foundation are represented by the inner circle in Fig. 7, and it is constantly pushed outwards by the growing demands of the earth's population.

The safe and just space for humanity, hence, has boundaries both in terms of an environmental ceiling and in terms of a social foundation. This means that beyond the planetary boundaries, in the outer layer of the "doughnut-shaped space", we face environmental degradation which endangers humanity and below the social boundary, which is the inner layer of the "doughnut-shaped space", we face deprivations that risk human well-being.

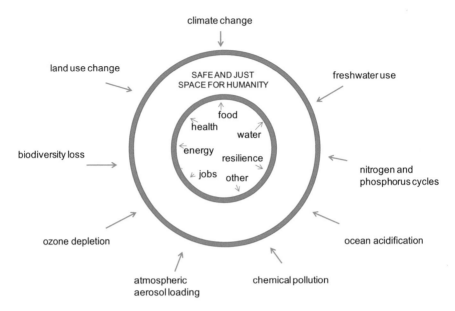

Fig. 7 An "environmentally safe and socially just operating space", modified from Dearing et al. (2014)

3.2.1 The Clothing Sector's Contribution to Human Needs

In terms of the ability of the clothing sector to fulfil the basic human needs, it is clear that clothes are an absolute necessity for survival. There are many different descriptions of what basic human needs are, but to select one that is often referred to in discussions on sustainable development, we begin with the framework on human-scale development presented by Manfred Max-Neef as early as 1987 (Boulanger et al. 2010). According to Max-Neef, there are nine basic human needs that are universal that are not changing over time and where one cannot replace the other. These are subsistence, protection, affection, understanding, participation, leisure, creation, identity and freedom. Max-Neef claims that all "cultures" make different efforts to address all of these needs but that they do it in very different ways. In Western society where consumerism is a strong characteristic of the culture, consumption of material goods is equated with needs fulfilment. When looking into the needs as expressed by Max-Neef, however, it soon becomes obvious that very few of them are directly connected to material goods. In fact, subsistence is the only one with a very strong connection—food, water, clothes, etc., are needed for survival. However, as we can establish that clothes will be necessary for human survival and that the clothing sector is therefore essential, we should also think about in what ways that clothes can help us fulfil other needs as we will have to use them anyway. Max-Neef urges us to reflect on how we can address the different needs in ways that are less materialistic and his framework provides a guide for this. Table 9 below presents the nine basic human needs and the relevance for the clothing sector.

Table 9 Basic human needs and relevance for the clothing sector

Need	Being (qualities)	Having (things)	Doing (actions)	Interacting (settings)	Examples of relevance for clothing sector
Subsistence	Physical and mental health	Food, shelter, work	Feed, clothes, rest, work	Living environment, social setting	The protective function of clothing is essential for survival
Protection	Care, adaptability, autonomy	Social security, health systems, work	Cooperate, plan, take care of, help	Social environment, dwelling	Protective function of clothing
Affection	Respect, sense of humour, generosity, sensuality	Friendships, family, relationships with nature	Share, take care of, make love, express emotions	Privacy, intimate spaces of togetherness	Clothing is used to gain affection in many ways
Understanding	Critical capacity, curiosity, intuition	Literature, teachers, policies, educational	Analyse, study, meditate, investigate	Schools, families, universities, communities	Clothing is used to signal belonging, mark cultural occasions[a], etc.
Participation	Receptiveness, dedication, sense of humour	Responsibilities, duties, work, rights	Cooperate, dissent, express opinions	Associations, parties, churches, neighbourhoods	Clothing is used to signal responsibilities, ideas, opinions, etc.
Leisure	Imagination, tranquillity, spontaneity	Games, parties, peace of mind	Daydream, remember, relax, have fun	Landscapes, intimate spaces, places to be alone	Clothing enables leisure in many ways
Creation	Imagination, boldness, inventiveness, curiosity	Abilities, skills, work, techniques	Invent, build, design, work, compose, interpret	Spaces for expression, workshops, audiences	Clothing is an important medium for creativity
Identity	Sense of belonging, self-esteem, consistency	Language, religions, work, customs, values, norms	Get to know oneself, grow, commit oneself	Places one belongs to, everyday settings	Clothing is used to signal belonging, identity, etc.
Freedom	Autonomy, passion, self-esteem, open-mindedness	Equal rights	Dissent, choose, run risks, develop awareness	Anywhere	Clothing is used to signal ideas, opinions, etc.

[a]weddings, funerals, festivities, etc.

Another description of basic human needs, which connects the needs with different ecosystem services, can be found in the Millennium Ecosystem Assessment (Millennium Ecosystem Assessment 2005). This scheme depicts the intensity of linkages between ecosystem services and human well-being. Basic material for good life, such as clothing, has a strong connection to both provisioning and regulating ecosystem services, and therefore, it is important to focus on them in a sustainability assessment. The scheme further investigates the potential for mediation by socio-economic factors.

3.3 Sustainability Targets in the Fashion Industry

So how can sustainability be defined in the context of the clothing industry? In this chapter, we have tried to identify and explore the impact of interventions aimed at improving the current sustainability performance of the Swedish fashion industry (the current sustainability performance is elaborated on in Sect. 2, and the interventions are further described in Sect. 4). Interventions can be applied on a garment level, a company level or a national sector level. Our starting point has been the garment level, or rather the function provided by the garment (the functional unit, in LCA terms), which was then scaled up to the functions provided by all garments purchased in one year in Sweden (as explained in Sect. 2). However, when evaluating the effectiveness of the interventions, i.e. their capacity to contribute towards reaching sustainability, we needed to define what we are aiming for.

3.3.1 Targets for Environmental Sustainability

To set environmental sustainability targets, we decided to try to define what the maximum impact per functional unit (on a garment level) can be in 2050 for the provision of the functional unit to be sustainable, taking into account that more functions will at that time have to be provided within the Swedish clothing sector (as the population of Sweden is expected to grow). That is, instead of defining what is possible to achieve based on today's state, we attempted to define what is necessary to achieve based on an envisioned future desired state and use that definition for steering further development. This can be seen as an example of back-casting (Robinson 1982). We deemed 2050 to be a reasonable time horizon for defining the desired state that requires a transformation of an entire industry sector, as it allows for the development and diffusion of interventions proposed today.

In defining the targets, we used the planetary boundaries framework and developed a procedure for translating what the framework tells us in terms of what impact reductions are needed in terms of various environmental impacts globally, into impact-reduction targets at the level of the functional unit. The procedure is presented in Sandin et al. (2015), where it is also applied in the context of the Swedish clothing consumption. Below, the content of Sandin et al. (2015) is summarised. Further, this section contains an update of the application of the procedure for Swedish clothing consumption compared to the paper and, in addition, the application of the procedure for the US and Chinese clothing consumption.

The procedure consists of four steps: (1) identifying the planetary boundaries quantified in the literature that correspond to the LCA impact category studied in the particular study; (2) interpreting what the identified planetary boundaries tell us about the need for reducing current global impacts until a chosen point in time (2050, in our case); (3) translating the global impact-reduction targets identified in step 2 into impact-reduction targets for the specific global market segment of concern (the clothing industry, in our case); and (4) translating the global impact-reduction targets identified in step 3 to impact-reduction targets per functional unit.

In applying the procedure in the context of Swedish clothing consumption, a number of challenges were identified in each step. In step 1, it was found to be far from a trivial exercise to match the planetary boundaries to LCA impact categories and to translate these to global targets for impact reduction until the year 2050. Table 10 lists the six planetary boundaries that have been quantified and that were deemed relevant in the context of the Swedish clothing sector, along with related LCA impact categories and the identified global targets for impact reduction. That impact categories are listed does not mean that the global impact-reduction targets are directly transferable to these impact categories. The specific matching of a target and an impact category depends, for example, on the control variable(s) of the planetary boundary and the characterisation method of the impact category. For the specific LCA context of Roos et al. (2015b), described in Sect. 2, we found that

Table 10 The outcome of steps 1 and 2 of the procedure in Sandin et al. (2015): the planetary boundaries that have been quantified and were deemed relevant in the context of the Swedish clothing sector, related LCA impact categories and the identified global targets for impact reduction

Planetary boundary	Related impact categories	Global target for impact reduction until 2050 as implied by the planetary boundaries framework (%)
Climate change	Climate change, non-renewable energy use	100
Interferences with the nitrogen cycle (part of the biogeochemical flows PB)	Eutrophication, marine eutrophication, terrestrial eutrophication, terrestrial acidification	59
Interferences with the phosphorus cycle (part of the biogeochemical flows PB)	Eutrophication, freshwater eutrophication	56
Freshwater use	Freshwater consumption	−54
Land system change	Land transformation (in particular transformation of forest land)	100
Changes in biosphere integrity	Land occupation (mid-point), land transformation (mid-point), biodiversity loss (endpoint)	99

the target for climate change was transferable to the LCA impact categories of climate change and non-renewable energy use and that the target for interference with the phosphorus cycle was transferable to the freshwater eutrophication impact category, as shown in Table 13.

Steps 3 and 4 of the procedure require value-based choices as the available space needs to be shared between market segments. For translating the global target to targets for the global clothing market segment (step 3), we employed three different ethical principles, in which the clothing market segment has (A) the right to cause the same share of the global impact in 2050 as of today, (B) the right to cause half the share of impact as it does today or (C) the right to cause twice the share of the impact it has today. This is a value-based choice related to how important one considers the clothing market segment to be in relation to other market segments.

For translating the impact-reduction targets for the global market segment to impact-reduction targets per functional unit (step 4), another highly value-based element, we employed four different ethical principles: (1) individual rights (i.e. the allowed impact space in 2050 should be split equally between all individuals in the world); (2) historical right of the regional market segment (i.e. the studied regional market segment, in our case the Swedish clothing sector, inherits the right to cause a certain share of the global impact space in 2050); (3) historical right of individuals in populations (similar to principle 2, but instead residents of the studied region (Sweden, in our case) inherit the right to cause a certain share of the global impact space in 2050); and (4) historical debt of individuals in populations (the opposite of principle 3; if residents of Sweden cause more impact than the global average today, future residents of Sweden should cause less impact than the global average).

The resulting impact-reduction targets for Swedish clothing consumption are shown in Table 11. These are updates of the results of Sandin et al. (2015), with more recent data for some of the parameters used in applying the ethical principles of step 4 (see Table 17 in the appendix). In Sandin et al. (2015), we also showed results for Nigerian clothing consumption, to test the influence of certain parameters which differ considerably between Sweden and Nigeria. In the present chapter, we instead show results also for the USA and China—important clothing markets which we believe broaden the scope of the chapter—see Tables 12 and 13 (background data for these calculations are shown in Table 17 in the appendix).

The results in Tables 11, 12 and 13 give an indication of the order of magnitude of the impact reduction needed for clothing consumption in Sweden, USA and China, in a society that respects the planetary boundaries. Specifically, the results indicate that considerable environmental impact reduction is necessary until 2050 for most types of environmental impacts. The necessary reductions span from eliminating all or nearly all impact (in impact categories such as climate change, fossil resource depletion, land transformation, biodiversity loss and eutrophication, for certain ethical principles and data assumptions applied), to even allowing some increase of impact (freshwater consumption and eutrophication, for certain ethical principles and data assumptions applied). As such, it gives some idea of what the Swedish, US and Chinese clothing sectors need to achieve to become sustainable. This can be a basis for setting sustainability targets in firms and sectors, in particular

Table 11 Impact-reduction targets for the clothing consumption in Sweden, applying the procedure outlined in Sandin et al. (2015)

Ethical principle (step 4)			Impact-reduction target per garment use [%]							
			Principle 1: individual rights		Principle 2: historical right of sectors	Principle 3: historical right of individuals of populations		Principle 4: historical debt of individuals of populations		
Global population in 2050 (step 4)			Low estimate	High estimate	Independent of global population	Low estimate	High estimate	Low estimate	High estimate
PB	Related impact categories	Approach (step 3)							
Climate change	Climate change, non-renewable energy use	A–C	100	100	100	100	100	100	100
Interferences with the nitrogen cycle	Eutrophication, marine eutrophication, terrestrial eutrophication, terrestrial acidification	A	72.3–87.0	77.6–88.3	66.2	71.5	77	81.7–96.0	85.2–96.7
		B	86.1–93.5	89.2–94.5	83.1	85.8	88.5	90.9–98.0	92.6–98.4
		C	44.5–73.9	55.3–79.0	34.2	43.1	54.1	63.4–91.9	70.5–93.5
Interferences with the phosphorus cycle	Eutrophication, freshwater eutrophication	A	82.1	85.6	64.7	69.4	75.4	92.9	94.3
		B	91.1	92.8	82.3	84.7	87.7	96.5	97.1
		C	64.2	71.1	32.5	38.9	54.1	85.8	88.6
Freshwater use	Freshwater consumption	A	−35.8	−9.48	−26.8	−6.95	13.8	−16.8	5.85
		B	32.2	45.3	36.6	46.5	55.9	41.6	52.9
		C	−172	−119	−154	−114	−72.5	−134	−88.3
Land system change	Land transformation	A–C	100	100	100	100	100	100	100
Change in biosphere integrity	Land occupation, land transformation, biodiversity loss	A	99.7	99.8	99.2	99.3	99.4	99.9	99.9
		B	99.9	99.9	99.6	99.7	99.7	100	100
		C	99.4	99.5	98.4	98.6	98.9	99.8	99.9

Targets are given per planetary boundary for the different approaches of step 3 of the procedure (A, B and C) and the different population developments and ethical principles of step 4. For the impact categories related to the PB of interferences with the nitrogen cycle, a range is given for ethical principles 1 and 4, spanned by the low and high estimates of the ratio of global to Swedish per capita impact of Sandin et al. (2015) [principles 2 and 3 are not influenced by this ratio (see Eqs. 3 and 4 in Sandin et al. (2015))]. A negative value implies that the impact is allowed to increase

Table 12 Impact-reduction targets for the clothing consumption in the USA, applying the procedure outlined in Sandin et al. (2015)

| Ethical principle (step 4) | | | Impact-reduction target per garment use [%] | | | | | | |
| Global population in 2050 (step 4) | | | Principle 1: individual rights | | Principle 2: historical right of sectors | Principle 3: historical right of individuals of populations | | Principle 4: historical debt of individuals of populations | |
PB	Related impact categories	Approach (step 3)	Low estimate	High estimate	Independent of global population	Low estimate	High estimate	Low estimate	High estimate
Climate change	Climate change, non-renewable energy use	A–C	100	100	100	100	100	100	100
Interferences with the nitrogen cycle	Eutrophication, marine eutrophication, terrestrial eutrophication, terrestrial acidification	A	84.9–89.3	87.8–91.4	50.5	58.2	66.3	92.0–96.0	93.5–96.8
		B	92.4–94.7	93.9–95.7	75.2	79.1	83.1	96.0–98.0	96.8–98.4
		C	69.7–78.6	75.6–82.7	0.9	16.4	32.6	84.0–92.0	87.1–93.6
Interferences with the phosphorus cycle	Eutrophication, freshwater eutrophication	A	85.7	88.5	46.8	55.1	63.8	93.4	94.7
		B	92.9	94.2	73.4	77.6	81.9	96.7	97.3
		C	71.4	77	-6.35	10.3	27.6	86.7	89.3
Freshwater use	Freshwater consumption	A	31.2	44.5	-86.1	-57	-26.6	56	64.5
		B	65.6	87.6	6.95	21.5	36.7	78	82.3
		C	-37.6	50.4	-272	63.6	-2.66	11.9	29
Land system change	Land transformation	A–C	100	100	100	100	100	100	100
Change in biosphere integrity	Land occupation, land transformation, biodiversity loss	A	99.7	99.8	98.8	99	99.2	99.9	99.9
		B	99.9	99.9	99.4	99.5	99.6	100	100
		C	99.5	99.6	97.6	98	98	99.8	99.8

Targets are given per planetary boundary for the different approaches of step 3 of the procedure (A, B and C) and the different population developments and ethical principles of step 4. For the impact categories related to the PB of interferences with the nitrogen cycle, a range is given for ethical principles 1 and 4, spanned by the low and high estimates of the ratio of global to USA per capita impact of Sandin et al. (2015) (principles 2 and 3 are not influenced by this ratio (see Eqs. 3 and 4 in Sandin et al. (2015)). A negative value implies that impact is allowed to increase

Table 13 Impact-reduction targets for the clothing consumption in China, applying the procedure outlined in Sandin et al. (2015)

Ethical principle (step 4)			Impact-reduction target per garment use [%]						
			Principle 1: individual rights		Principle 2: historical right of sectors	Principle 3: historical right of individuals of populations		Principle 4: historical debt of individuals of populations	
Global population in 2050 (step 4)			Low estimate	High estimate	Independent of global population	Low estimate	High estimate	Low estimate	High estimate
PB	Related impact categories	Approach (step 3)							
		A–C	100	100	100	100	100	100	100
Climate change	Climate change, non-renewable energy use								
Interferences with the nitrogen cycle	Eutrophication, marine eutrophication, terrestrial eutrophication, terrestrial acidification	A	51.7–77.9	61.1–82.2	59.8	66.1	72.7	18.6–86.2	34.4–88.9
		B	75.9–89.0	80.5–91.1	79.9	83.1	86.3	59.3–93.1	67.2–94.5
		C	3.46–55.9	22.1–64.5	19.7	32.2	45.3	–62.8–72.5	eutrophication, Terrestrial 31.3–77.8
Interferences with the phosphorus cycle	Eutrophication, freshwater eutrophication	A	67.2	73.6	56.9	63.6	70.7	71.6	77.1
		B	83.6	86.8	78.5	81.8	85.3	85.8	88.6
		C	34.4	47.1	13.8	27.3	41.3	43.2	54.2
Freshwater use	Freshwater consumption	A	–85.7	–49.7	–50.9	–27.3	–2.67	–160	–109.5
		B	7.18	9.77	24.6	36.4	48.7	–29.9	–4.76
		C	–271	–261	–202	–155	–105	–420	–319
Land system change	Land transformation	A–C	100	100	100	100	100	100	100
Change in biosphere integrity	Land occupation, land transformation, biodiversity loss	A	99.4	99.5	100	99.2	99.3	99.5	99.6
		B	99.7	99.7	100	99.6	99.7	99.8	99.8
		C	98.7	99	98	98.4	98.7	99	99.2

Targets are given per planetary boundary for the different approaches of step 3 of the procedure (A, B and C) and the different population developments and ethical principles of step 4. For the impact categories related to the PB of interferences with the nitrogen cycle, a range is given for ethical principles 1 and 4, spanned by the low and high estimates of the ratio of global to China per capita impact of Sandin et al. (2015) (principles 2 and 3 are not influenced by this ratio (see Eqs. 3 and 4 in Sandin et al. (2015))). A negative value implies that impact is allowed to increase in setting targets for reducing the impact per garment, which is a tangible and common level when working with interventions for impact reduction (as further elaborated on in Sect. 4). The procedure cannot yet, however, provide input to reduction targets for all the important environmental challenges of the clothing sector, for example toxicity issues. This is both because all the nine planetary boundaries have not yet been quantified and because there is a need to further research the connection between the planetary boundaries and the metrics at which environmental impact is quantified in firms and sectors (e.g. specific metrics of LCA impact categories).

Furthermore, the results in Tables 11, 12 and 13 indicate that impact-reduction targets, derived using the planetary boundary framework, depend heavily on the chosen ethical principles in steps 3 and 4, the estimated global population in 2050 [we tested the low and high estimates of UN (2015a)] and the geographical context. This underlines the inherent value-based choices one by necessity has to make when dividing a finite impact space between different entities having a claim for that space, as well as the inherent uncertainties involved when speculating about the future (in this case, the global and national populations in 2050).

It is important to once again stress that connecting the planetary boundaries with certain impact categories is difficult, and it can be questioned if some global planetary boundaries (e.g. the one for freshwater use) can (yet) guide impact-reduction efforts and target-setting, without consideration of local or regional geographical parameters, such as the water stress in specific water basins.

In the LCA context described in Sect. 2, only four impact-reduction targets were deemed feasible to connect to specific impact categories. In doing this, we chose to use ethical principle A in step 3 and ethical principle 1 in step 4—which are, albeit, value-based choices, perhaps the least controversial ones. Employing the same interpretation for the results in Tables 11, 12 and 13 yields the results of Table 14. The freshwater consumption "target" is not derived using the procedure of Sandin et al. (2015), but is the guidance on maximum acceptable water withdrawal at the basin level provided by Steffen et al. (2015).

Table 14 Suggested impact-reduction targets in the LCA impact categories of the study described in Sect. 2, based on the maximum acceptable environmental impact for clothing consumption in Sweden, USA and China for year 2050 according to the planetary boundaries framework, reworked from Roos et al. (2016)

Impact categories	Impact-reduction target per garment use for 2050
Climate change	100 % compared to current level, i.e. "climate neutral"
Freshwater eutrophication	Sweden: 82–86 % compared to current level USA: 86–89 % compared to current level China: 67–74 % compared to current level
Freshwater consumption	Blue water withdrawal as % of mean monthly river flow: for low-flow months: 25 %; for intermediate-flow months: 30 %; for high-flow months: 55 %
Non-renewable energy use	100 % compared to current level

3.3.2 Targets for Social Sustainability

To set social sustainability targets for the clothing sector, we used a set of 11 social targets based on an analysis of governmental submissions to the United Nations Rio +20 Conference in 2012 (Leach et al. 2013), just as was done by Raworth in 2013 (Raworth 2013). After that work of Raworth, in which she also identified the share of the population that experienced deprivations in terms of selected indicators for each of the indicator fields of the social foundation, further discussions have led to the agreement of the SDGs. Table 15 contains a list of the 11 social targets and when there is a match, the connected SDG. Note that if the SDGs are explored in another context, there will be targets that connect to other social sustainability targets.

As for the environmental aspects, we want to find a way to relate these sustainability targets to the social impacts that can be assessed in an SLCA. Table 15 also shows how the indicators used for social hotspot identification of clothing consumption in Sweden, China and USA in Sect. 2.2 connect to the social sustainability targets and the SDGs. For several reasons, it is more difficult and perhaps less relevant to use a strictly quantitative approach in this case. The most important reason in this case is that the social hotspot identification will only reveal hotspots where considerable risks of deprivations will be found in the value chain. Therefore, we settled in our study for a qualitative discussion on how different interventions can ease or exacerbate such deprivations.

However, setting quantitative social sustainability targets can be possible. In Raworth (2013), indicators are suggested for most of the 11 social sustainability targets listed in Table 15, and each of the SDGs has a list of more specific targets to be achieved. Taking adequate income as an example, Raworth (2013) identifies the share of the population living below $1.25 (purchasing power parity) per day to be 21 % in 2005 and this is clearly a measure of the extent of deprivations and implying that the target should be that no person is living under this limit. Under SDG 1, there is also a more detailed description of what this might mean in terms of different targets. In our case, Table 15 should be seen as an attempt to identify important areas that connect to different social foundations. For each of the areas where there is a match between the social foundation and indicators that are assessed in SLCA, relevant targets will have to be discussed in light of the specific study at hand. In some cases, the mapping of the targets, goals and indicators is relatively weak. In particular, neither Leach et al. (2013) nor Raworth (2013) are completely clear on what they mean by "resilience to shocks" though the latter refers to poverty as the key problem affecting resilience, which is why we might connect it to a SDG concerned with inclusive and sustainable cities. On the other hand, the indicators in the last column are more to do with workplace safety, and while "safe" is one of the descriptors of that SDG, one might equally connect the indicators we have to the social sustainability target of "decent job". This is an example of the challenges faced by analysts attempting to associate goals objectives and indicators for socially sustainable development.

The final step in our process that is needed to be able to evaluate interventions is the translation of global targets to the market segment that is addressed in the

Table 15 Social sustainability targets according to the Rio+20 conference, as identified by Leach et al. (2013), related SDGs (United Nations 2015b) and equivalent social indicators used in social hotspot identification

Social sustainability target	Related SDG	Equivalent social indicators
Food security	2: End hunger, achieve food security and improved nutrition and promote sustainable agriculture	Not identified
Adequate income	1: End poverty in all its forms everywhere	Risk of wages being under 2 USD per day Risk of a sector average wage being lower than country's minimum wage
Improved water and sanitation	6: Ensure availability and sustainable management of water and sanitation for all	Not identified
Adequate health care	3: Ensure healthy lives and promote well-being for all at all ages	Not identified
Provide education	4: Ensure inclusive and equitable quality education and promote lifelong learning opportunities for all	Risk of child labour in sector
Decent job	–	Risk that country has not ratified ILO conventions by sector Risk that country does not provide adequate labour laws by sector Risk that a country lacks or does not enforce collective bargaining rights
Access to modern energy services	7: Ensure access to affordable, reliable, sustainable and clean energy for al	Not identified
Resilience to shocks	11: Make cities and human settlements inclusive, safe, resilient and sustainable	Risk of fatal injury by sector Risk of non-fatal injury by sector Overall risk of loss of life years by exposure to carcinogens in occupation Overall risk of loss of life years by airborne particulates in occupation
Gender equality	5: Achieve gender equality and empower all women and girls	Risk of gender inequality by sector based on representation in the workforce
Social equity	10: Reduce inequality within and among countries	Not identified
Having political voice	16: Promote peaceful and inclusive societies for sustainable development, provide access to justice for all and build effective, accountable and inclusive institutions at all levels	Not identified

specific study, in our case, the Swedish clothing sector. As many social targets are not set on a global scale but rather relate to the deprivations of individuals, there is no "operating space" that needs to be shared between countries of market segments in the same way as for environmental stressors. As an example, in Roos et al. (2016), we suggested that meeting minimum wages for supply chain employees is an appropriate goal for 2020 for the social issue of inadequate wage (wage under 2 USD). The minimum wages currently vary between countries. As an example, for Bangladesh where the minimum wage is set by the National Minimum Wage Board, it is 39 USD per month (ILO 2013). For 2050, we suggested meeting living wages as a goal (Roos et al. 2016). Several organisations set living wages in different countries. One example that is relevant for the textile sector is Asia Floor Wage, setting 260 Euro (approximately 280 USD) per month as the living wage for Bangladesh (Asia Floor Wage 2013).

4 Evaluation of Interventions

This section contains an evaluation of the potentially reduced detrimental environmental and social impacts of clothing, from different types of interventions. The evaluation of environmental impact reduction focuses on climate impact. The social impact reduction is evaluated qualitatively.

4.1 Interventions for Environmental Impact Reduction

There are many proposals for interventions that aim to reduce the environmental burden of textiles (Hasanbeigi and Price 2015). Some of the most commonly suggested interventions are textile recycling (Nordic Council of Ministers 2015), non-fluorinated water-repellent treatment (Schellenberger et al. 2015), forest-based fibres (Shen et al. 2010), collaborative consumption (Leismann et al. 2013), changing from wet to dry technologies (Euratex 2016) and reduced washing temperature (Muthu 2015a, b). However, the potential of such interventions to reduce the environmental impact of an entire national textile sector is seldom evaluated quantitatively.

In Roos et al. (2016), ten interventions for environmental impact reduction were evaluated with LCA. The studied interventions included a change from traditional, linear business models to collaborative consumption business models [more specifically, clothing libraries; see more details in Zamani et al. (2016b)], increased material recycling, a change from cotton to forest-based fibres and various behavioural changes of consumers, as shown in Table 16.

Table 16 Ten interventions for reducing the climate impact of clothing consumption (Roos et al. 2016)

No.	Scenario name	Scenario description
1	Offline collaborative consumption	40 % of the Swedish clothing consumption is turned into collaborative consumption in physical stores, doubling the garment service life
2	Online collaborative consumption	40 % of the Swedish clothing consumption is turned into collaborative consumption on the internet, doubling the garment service life
3	Polyester recycling	100 % of the virgin polyester in garments is replaced by chemically recycled polyester (repolymerisation)
4	Mechanical cotton recycling	15 % of the virgin cotton in garments is replaced by mechanically recycled cotton
5	Cotton recycled to lyocell	100 % of the virgin cotton in garments is replaced by chemically recycled cotton
6	Forest-based lyocell	100 % of the virgin cotton in garments is replaced by lyocell from virgin forest-based material
7	Long service life	The consumer uses garments twice as many times, i.e. doubling the garment service life
8	Renewable energy	Only renewable energy is used throughout the life cycle of garments (production, transportation, laundry, etc.)
9	Energy efficiency	20 % less energy is used in garment production
10	Human-powered transport	The consumers go by foot or on bike to the store

In the case of collaborative consumption (interventions 1 and 2 in Table 16), the sustainability benefit derives from an increased service life of garments and the associated reduction of garment production, while the number of garments a given consumer uses is not reduced but increased. In other words, from the perspective of the garment, fashion is slowed down (each garment is used more times before disposal), and from the perspective of the user, fashion is speeded up (each user wears more garments in a given time period). Increasing the service life is possible if garments are not used for their technical service life, which is the case in Sweden. Increased service life is probably not, however, feasible for all garments, and for the purpose of illustration, an ad hoc chosen subset of the clothing consumption is investigated—the cotton-based fraction, which represents roughly 40 % of the total consumption. There are three scenarios with material level recycling of polyester and cotton (interventions 3–5), polyester making up 36 % of the total flow and cotton 40 % of the total flow. It should be noted that increased service life (intervention 7) implies a reduction of money spent on fashion consumption (in contrast to the collaborative consumption scenarios, interventions 1 and 2) and that the financial savings associated with this intervention can give rise to more complex outcomes including the so-called rebound effects.

4.2   Climate Change Reduction of Selected Interventions

Figure 8 below shows the potential climate savings from the different interventions compared to the base case.

All the interventions except one show climate impact-reduction potential. In the case of offline collaborative consumption, there is a risk that the climate impact in fact increases due to consumers travelling more often to and from the store (under certain transportation assumptions). Longer service life and change to renewable energy sources show the largest climate impact-reduction potential. The interventions can also be combined in most cases, i.e. the implementation of one intervention does not prevent the implementation of some other interventions.

Only results for climate change are shown in Fig. 8, although impact from the clothing life cycle on water consumption, emissions of hazardous substances and land use are also important aspects (Roos et al. 2015a; Sandin et al. 2013). For results on other impact categories, the reader is referred to Roos et al. (2015b).

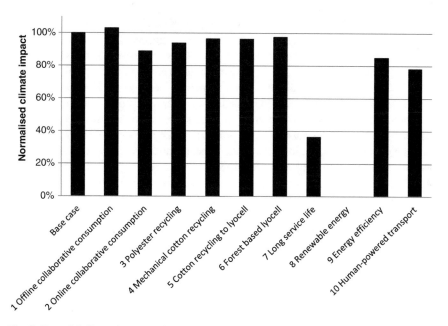

Fig. 8 Potential climate impact of the baseline model of Swedish clothing consumption as well as baseline model combined with various interventions for impact reduction. Results are normalised to the baseline model

4.3   Interventions for Social Impact Reduction and Their Effect

With the qualitative approach selected in this study for social impacts, a discussion can be made around how different interventions aimed at reducing deprivations of the social foundations of well-being can be effective or not. We do this below primarily for the example of inadequate wages.

In this study, an inadequate wage (wage under 2 USD) appeared to be one of the high-risk social issues in relation to clothing consumption in both China and Sweden (see Sect. 2.2). Several actions focused on the reduction of negative social impacts have been initiated by industry for improving the wage conditions in the fashion sector. An example of this is the Fair Living Wage roadmap, a programme introduced by the clothing retailer H&M, that has initiated a number of strategies for improving workers' living conditions. Some examples are as follows: develop purchasing practices to support suppliers in paying a fair living wage, pilot a fair wage method in some selected role model factories in Cambodia and Bangladesh, implement improved pay structures at all strategic supplier factories, implement plans to strategically engage with governments on wage issues and form a strategic partnership with the International Labour Organization (ILO) (H&M 2014). At this stage, it is too early to detect in a quantitative way impacts of interventions such as these. However, it seems clear that they do address this aspect of the social foundation of human well-being and that there is a good chance that they will lead to an improvement in living conditions for many people (Roos et al. 2016).

Similar discussions can be held in relation to other indicators related to the social foundations. Furthermore, other interventions, such as the ones listed in Table 16 that were suggested because of their potential to decrease the climate impact of clothing, can be discussed in terms of their social impact. With the I/O-based social hotspot identification that was performed, described in Sect. 2.2, in general, reducing the need for import of textiles from countries in which high risks of social deprivations were identified will also decrease the risks as they were defined as social hotspots. However, in reality, when this is done in larger scale, market effects might lead to completely different consequences as the value chains are reorganised to accommodate for the large change. As an example, if consumers avoid Bangladeshi clothing products on account of concerns about labour abuse, this might depress the price of such products from that country and drive the wages and conditions of Bangladeshi textile workers down further. For this reason, it would be a rather facile form of analysis to merely shift production from one country to another in a static I/O model and concern oneself with industry in the country to which the production is moved. Therefore, more dynamic modelling and social impact research are needed into how increased circularity of textile flows will change the fashion industry. Further, a comparison between updates and databases such as the SHDB may be useful for monitoring changes in relation to risk levels for social issues after implementing different social (or other) interventions. Nevertheless, one may expect the social risk levels in relation to inadequate wages will be reduced by implementing such actions.

5 Discussion

The idea of a sustainable fashion industry may seem paradoxical, as stated in the introduction to this chapter. Fundamentally, we need *clothing* for protection from variations in the weather, which is distinct from *fashion* that is usually seen as driving cosmetic wants and consumption. It is possible, however, to think of a human need for personal communication and that this need has been expressed via our clothing choices over the recorded history of humankind, long before current environmental concerns about the negative impacts of fashion emerged. Issues which challenge the clothing industry today, such as the excess use of limited resources, the pollutants the industry releases and the working conditions within it, are consequences of the modern scale and methods of the industry, so sustainable fashion ought to be possible, if we can decouple it from its current issues.

Working conditions are another challenge for the sustainability of the fashion industry. The globalisation of the fashion industry over the last 30 years has perhaps lifted many people in Asia from poverty, but it has also created concerns about the abuse of labour in countries that do not have strong labour representation. A systemic aspect of the problem is that the supply chains are now so globalised and complex that it can be difficult for managers, for example those in a European clothing retailer, to know precisely where the garments they sell are being produced. The long supply chain between buyer, contractor and many subcontractors isolates management from distant workplaces, so that even if consumers demand better conditions for garment production, their implementation is challenging.

5.1 Growing Use of LCA in the Clothing Sector

The use of LCA in the clothing industry, by authorities, by researchers and by the industry itself, is growing rapidly from a low start and is increasingly used in environmental declarations, procurement guidelines, product design and regulatory work, all of which potentially impact upon the entire life cycle of a textile product (Curwen et al. 2013).

In the EU Integrated Product Policy work, LCA was identified as the "best framework for assessing the potential environmental impacts of products currently available" (European Commission 2013). The European "Ecodesign Directive" (European Commission 2009) as well as the European Commission initiative for Product Environmental Footprint (PEF) (European Commission 2013) is based on LCA and is currently in the pilot phase where textiles are one of the pilot cases.

The SAC (2016) also encourages LCA-based environmental product declarations (EPD) of textile products and has developing guidance material for how to create Product Category Rule (PCR) (Schenk 2013).

5.2 LCA as an Quantitative Approach to Textile Sustainability

LCA is today one of several available complementary tools for managing sustainability in the textile industry (Roos et al. 2015a), with a singularity in that it includes a quantitative evaluation of multiple environmental issues along the whole product life cycle. The Cradle to Cradle® (C2C®) methodology is another commonly applied framework for sustainable design of textile products (McDonough and Braungart 2002). LCA differs from C2C® in that it is a quantitative and holistic methodology and is independent of commercial interests, stated Bor et al. (2011). Other textile companies choose to follow schemes such as Bluesign® (BLUESIGN® 2013) to manage environmental impacts in the supply chain, the Business Social Compliance Initiative (BSCI 2013) for social sustainability or OEKO-TEX® Standard 100 (OEKO-TEX® Association 2013) to ascertain the absence of hazardous chemicals in the textile chain. The Higg Index is a tool under development from the Sustainable Apparel Coalition (SAC) (2016) for assessing the sustainability of brands, facilities and products and has in common with the Chemicals Management Framework from the Chemicals Management Working Group (Outdoor Industry Association 2014) that it is based on the evaluation of management procedures, in addition to being based on quantitative data on environmental and social impact (partly based on LCA methodology). LCA stands out as a tool that can give quantitative answers considering multiple environmental issues along the whole life cycle of alternative products, technologies and management procedures to designers, purchasers and consumers (Roos et al. 2015a).

5.3 LCA on Industrial Sector Level

Roos et al. (2016) showed how LCA can be used not just for product evaluation but potentially for sector-scale assessments. Some insights from the industry sector perspective highlighted in this chapter are as follows:

- Previous studies are based on expected technical life length, not statistics of real user behaviour. This strongly influences the relative importance of different life cycle phases.
- The importance of laundry practices (e.g. washing temperature) depends on a geographical and demographical context.

- In Sweden, the consumption of new clothes is high and the climate impact of electricity is low, rendering the washing temperature a low importance for Sweden (3 % of total life cycle impact).
- In the USA, the consumption of new clothes is high and the climate impact of electricity is higher, rendering the washing temperature a medium importance for USA (27 % of total life cycle impact).
- In China, the consumption of new clothes is moderate and the climate impact of electricity is high, rendering the washing temperature a high importance for China (68–75 % of total life cycle impact).

• Consumer transport might be more important than what has previously been understood, and this must be considered when developing, e.g., clothing libraries and other sharing economy business models.
• To be able to evaluate the potential of interventions, and the need for more radical system changes, we need to know how far the clothing industry is from sustainability in environmental and social terms.

- The planetary boundaries framework can be used to increase this understanding—a procedure for this is described in this chapter—but not yet for all the relevant LCA impact categories. Also, evaluating sustainability in absolute terms requires value-based decision on how to allocate allowed impact space between industry sectors and nations.
- Among others, the proposed procedure indicates that to respect the planetary boundaries, the climate impact and non-renewable energy use of the clothing sector must be eliminated by 2050 and that freshwater eutrophication must be reduced by about 67–74 % in China, 82–86 % in Sweden and 86–89 % in the USA.

• Studying the climate benefits of 10 interventions aimed at impact reduction showed that prolonging garment service lives and switching to renewable energy use are the most effective interventions.
• The available SLCA data are on country and economic sector level. Social metrics need thus to be refined in order to measure the influence of company level interventions on reaching the targets.

5.4 Concluding Remarks

To answer the title question of this chapter, will clothing be sustainable, there are clearly challenges. Section 1 describes the systemic challenges for sustainable fashion, current initiatives and the different actors involved. Section 2 elaborates on the technical challenges in assessing the status of sustainability in the clothing industry today. The complexity of the textile supply chain has led to that many simplifications and assumptions have been made, which have in some cases been used in a broader context than what they originally were intended for, i.e. being assumed to have a higher generality and representativeness than they actually have (described in 5.3). Section 3 is an attempt to describe the state of "sustainability", to give us a notion of what we strive to achieve. If we know the state of the environmental and social impacts of the clothing industry today, and we know how the state of "sustainability" could look like, we can suddenly see whether we are near sustainability or far from it. Section 4 evaluates quantitatively the climate savings that would be the result of some commonly proposed interventions.

Considering climate change, the results show that only one of the proposed interventions—replacing all fossil energy used in the product life cycle with renewable energy—leads to the presented goal for sustainability. However, considering the unlikeliness of fully abandoning non-renewable energy sources before 2050, and that sustainability is about so much more than climate change, it is clear that several interventions need to be implemented. The industry, policy makers, consumers and citizens must thus collaborate to make clothing sustainable.

Appendix See Table 17.

Table 17 Data for applying the ethical principles in step 4 of the procedure of Sandin et al. (2015), in the context of the Swedish, USA and Chinese clothing sectors, as described in Sect. 3.3

	Data			References
Global population				
Current (P_{GloCur})	7.35 billion			United Nations (2015a)
Low estimate 2050 (P_{GloFut})	8.71 billion			United Nations (2015a)
High estimate 2050 (P_{GloFut})	10.8 billion			United Nations (2015a)
National population	*Sweden*	*US*	*China*	
Current (P_{RegCur})	9.78 million	322 million	1380 million	United Nations (2015a)
Estimate 2050 (P_{RegFut})	11.9 million	389 million	1350 million	United Nations (2015a)
Ratio of global to national per capita impact (I_{Glo}/I_{Reg})	*Sweden*	*US*	*China*	
Interferences with the global nitrogen cycle	0.38–0.80	3.57–8.66	0.31–0.44	Low figure for Sweden and US, high figure for China: (Lenzen et al. 2013); High figure for Sweden and US, low figure for China: (Mekonnen and Hoekstra 2011)
Interferences with the global phosphorus cycle	0.48	0.39	0.88	Metson et al. (2012)
Global freshwater use	1.04	0.63	1.43	Mekonnen and Hoekstra (2011)
Changes in biosphere integrity	0.36	0.32	0.76	GFN (2016)

The abbreviations in parenthesis in the first column correspond to the terms used in Eqs. 2–5 of Sandin et al. (2015)

References

Allwood, J. M., Laursen, S. E., Rodriguez, C. M., & Bocken, N. M. P. (2006). Well dressed? The present and future sustainability of clothing and textiles in the United Kingdom, University of Cambridge Institute for Manufacturing. University of Cambridge, Institute for Manufacturing, Cambridge.

Asia Floor Wage. (2013). Living wage versus Minimum wage [WWW Document]. http://asia.floorwage.org/living-wage-versus-minimum-wage. Accessed 29 October 2015.

Baumann, H., & Tillman, A.-M. (2004). *The hitchhiker's guide to LCA*. Lund, Sweden: Studentlitteratur.

Benoit-Norris, C., Cavan, D. A., & Norris, G. (2012). Identifying social impacts in product supply chains: Overview and application of the social hotspot database. *Sustainability, 4*, 1946–1965. doi:10.3390/su4091946

Beton, A., Dias, D., Farrant, L., Gibon, T., Le Guern, Y., Desaxce, M., et al. (2014). Environmental Improvement Potential of textiles (IMPRO Textiles). European Union, Luxembourg: Publications Office of the European Union, 2014. doi:10.2791/52624

Bevilacqua, M., Ciarapica, F. E., Giacchetta, G., & Marchetti, B. (2011). A carbon footprint analysis in the textile supply chain. *International Journal of Sustainable Engineering, 4*, 24–36. doi:10.1080/19397038.2010.502582

BIO. (2007). Analyse de Cycle de Vie comparée d'une chemise en lin et d'une chemise en coton.

BLUESIGN®. (2013). BLUESIGN® [WWW Document]. http://www.bluesign.com/. Accessed 24 August 2013.

Bor, A.-M., Hansen, K., Goedkoop, M., Rivière, A., Alvarado, C., & van den Wittenboer, W. (2011). *Usability of life cycle assessment for cradle to cradle purposes*. Dutch Ministry of Economic Affairs, Utrecht, The Netherlands: NL Agency.

Boulanger, M., Cruz, I., Frühmann, J., Leßmann, O., & Max-Neef, M. (2010). *Sustainable development: Capabilities, needs, and well-being*. London: Routledge.

Brander, A. M., Sood, A., Wylie, C., Haughton, A., Lovell, J., Reviewers, I., et al. (2011). Technical Paper|Electricity-specific emission factors for grid electricity, Ecometrica.

BSCI. (2013). BSCI [WWW Document]. http://www.bsci-intl.org/. Accessed 24 August 2013.

Carbonaro, S., & Goldsmith, D. (2015). Branding sustainability. In K. Fletcher, M. Tham (Eds.), Handbook of sustainability and fashion. Routledge.

Clancy, G., Fröling, M., & Peters, G. (2015). Ecolabels as drivers of clothing design. *Journal of Cleaner Production, 99*, 345–353.

Collins, M., & Aumônier, S. (2002). *Streamlined life cycle assessment of two Marks & Spencer plc apparel products*. Oxford, UK: Environmental Resources Management Ltd.

Curwen, L. G., Park, J., & Sarkar, A. K. (2013). Challenges and solutions of sustainable apparel product development: A case study of Eileen Fisher. *Clothing and Textiles Research Journal, 31*, 32–47. doi:10.1177/0887302X12472724

Dearing, J. A., Wang, R., Zhang, K., Dyke, J. G., Haberl, H., Hossain, M. S., et al. (2014). Safe and just operating spaces for regional social-ecological systems. *Global Environmental Change, 28*, 227–238. doi:10.1016/j.gloenvcha.2014.06.012

Eerkes-Medrano, D., Thompson, R., & Aldridge, D. (2015). Microplastics in freshwater systems: A review of the emerging threats, identification of knowledge gaps and prioritisation of research needs. *Water Research, 75*, 63–82.

Ekener-Petersen, E., Höglund, J., & Finnveden, G. (2014). Screening potential social impacts of fossil fuels and biofuels for vehicles. *Energy Policy, 73*, 416–426. doi:10.1016/j.enpol.2014.05.034

Euratex. (2016). European Technology Platform for the Future of Textiles and Clothing [WWW Document]. http://www.textile-platform.eu/. Accessed 29 March 2016.

European Commission. (2014). Product Environmental Footprint (PEF) [WWW Document]. http://ec.europa.eu/environment/eussd/smgp/ef_pilots.htm. Accessed 1 June 2014.

European Commission. (2013). Commission Recommendation of 9 April 2013 on the use of common methods to measure and communicate the life cycle environmental performance of products and organisations (2013/179/EU).

European Commission. (2009). Directive 2009/125/EC of the European Parliament and of the Council of 21 October 2009 establishing a framework for the setting of ecodesign requirements for energy-related products. *Official Journal of the European Union, 285*, 10–35.

Faberi, S. (2007). Domestic Washing Machines and Dishwashers. Preparatory studies for Eco-design requirements of EuP. 2007.

Font Vivanco, D., Kemp, R., & Van Der Voet, E. (2015). The relativity of eco-innovation: Environmental rebound effects from past transport innovations in Europe. *Journal of Cleaner Production, 101*, 5380.

Garrido, S., Parent, J., Beaulieu, L., & Reveret, J.-P. (2015). A literature review of type I SLCA—making the logic underlying methodological choices explicit. *The International Journal of Life Cycle Assessment.* (in press).

GFN. (2016). At a glance. Global Footprint Network. [WWW Document]. www.footprintnetwork. org/en/index.php/GFN/page/at_a_glance/. Accessed 1 March 2016).

Granello, S., Jönbrink, A. K., Roos, S., Johansson, T., & Granberg, H. (2015). Consumer behaviour on washing.

Gwozdz, W., Netter, S., Bjartmarz, T., & Reisch, L. A. (2013). Report on Survey Results on Fashion Consumption and Sustainability among Young Swedes. Copenhagen.

H&M. (2014). Conscious action, sustainability report 2014. Stockholm, Sweden.

Hasanbeigi, A., & Price, L. (2015). A technical review of emerging technologies for energy and water efficiency and pollution reduction in the textile industry. *Journal of Cleaner Production, 95*, 30–44. doi:10.1016/j.jclepro.2015.02.079

Hellweg, S., & Milà i Canals, L. (2014). Emerging approaches, challenges and opportunities in life cycle assessment. *Science, 344*, 1109–1113. doi:10.1126/science.1248361

Holmquist, H., Schellenberger, S., van der Veen, I., Peters, G. M., Leonards, P. E. G., & Cousins, I. (2016). Properties, performance and associated hazards of state-of-the-art durable water repellent (DWR) chemistry for textile finishing. *Environment International, 91*, 251–264.

ILO. (2013). Minimum wage chart: Bangladesh, Cambodia & Viet Nam [WWW Document]. http://www.ilo.org/asia/WCMS_223988/lang–en/index.htm. Accessed 29 October 2015.

IPCC. (2013). Climate change 2013: The physical science basis. Working group I contribution to the fifth assessment report of the Intergovernmental Panel on Climate Change. In T. F. Stocker, D. Qin, G.-K. Plattner, M. Tignor, S. K. Allen, J. Boschung et al. Cambridge, United Kingdom and New York, NY, USA.

ISO. (2006). Environmental management—life cycle assessment—principals and framework. International Standard ISO 14040. International Organization for Standardization—ISO, Geneva, Switzerland.

Krozer, A., Björk, A., Hanning, A.-C., Wendel, A., Magnusson, E., Persson, F., Holmberg, K., & Jelse, K. (2011). Clean development and demonstration—Sustainable domestic washing—s'wash. Final report. IVL Swedish Environmental Research Institute, Stockholm, Sweden.

Larsson, J. (Ed.). (2015). Hållbara konsumtionsmönster Analyser av maten, flyget och den totala konsumtionens klimatpåverkan idag och 2050. Rapport 6653. Stockholm.

Laursen, S. E., Hansen, J., Knudsen, H. H., Wenzel, H., Larsen, H. F., & Kristensen, F. M. (2007). EDIPTEX—Environmental assessment of textiles.

Leach, M., Raworth, K., & Rockström, J. (2013). Between social and planetary boundaries: Navigating pathways in the safe and just space for humanity. In Social Science World Social Science. UNESCO, Paris.

Lefèvre, C. (2009). Ecodesign of Laundry Dryers. Preparatory studies for Ecodesign requirements Final Report March 2009. PricewaterhouseCoopers Advisory, France.

Leismann, K., Schmidt, M., Rohn, H., & Baedeker, C. (2013). Collaborative consumption: Towards a resource-saving consumption culture. *Resources, 2*, 184–203.

Lenzen, M., Daniel, M., Kanemoto, K., & Geschenke, A. (2013). Building Eora: A global multi-region input-output database at high country and sector resolution. *Economic Systems Research, 25*, 20–59.

Levi Strauss & Co. (2015). The life cycle of a jean. Understanding the environmental impact of a pair of Levi's® 501® jeans.

McDonough, W., & Braungart, M. (2002). *Remarking the way we make things: Cradle to cradle* (1st ed.). New York: North Point Press.

Mekonnen, M. M., & Hoekstra, A. Y. (2011). National water footprint accounts: the green, blue and grey water footprints of production and consumption. Volume 2: appendices. In *Value of Water Research Report Series No. 50. UNESCO-IHE Institute for Water Education*, Delpht, The Netherlands.

Metson, G., Bennett, E., & Elser, J. (2012). The role of diet in phosphorus demand. *Environmental Research Letters, 7*. doi:10.1088/1748-9326/7/4/044043

Millennium Ecosystem Assessment. (2005). *Ecosystems and human well-being: biodiversity synthesis*. Washington DC: World Resources Institute.

Mistra Future Fashion c/o SP. (2015). Mistra Future Fashion [WWW Document]. http://www.mistrafuturefashion.com Accessed 1 December 2015.

Muthu, S. (2015a). Environmental impacts of the use phase of the clothing life cycle. In S. Muthu (Ed.), *Handbook of LCA of textiles and clothing*. Cambridge: Woodhead Publishing/Elsevier.

Muthu, S. S. (2015b). *Social life cycle assessment. An Insight*. Singapore: Springer. doi:10.1007/978-981-287-296-8

Nordic Council of Ministers. (2015). *Well dressed in a clean environment*. Copenhagen, Denmark: Nordic action plan for sustainable fashion and textiles.

OEKO-TEX® Association. (2013). OEKO-TEX® Standard 100 [WWW Document]. https://www.oeko-tex.com. Accessed 24 August 2013.

Outdoor Industry Association. (2014). Chemicals Management Working Group (CMWG) [WWW Document]. http://outdoorindustry.org/responsibility/chemicals/index.html. Accessed 1 June 2014.

Peters, G. M., Granberg, H., & Sweet, S. (2014). The role of science and technology in sustainable fashion. In K. Fletcher & M. Tham (Eds.), *Routlegde handbook of sustainability and fashion*. London: Taylor & Francis.

Peters, G. M., Sack, F., Lenzen, M., Lundie, S., & Gallego, B. (2008). Towards a deeper and broader ecological footprint. *Proceedings of the Institution of Civil Engineers-Engineering Sustainability, 161*, 31–37. doi:10.1680/ensu.2008.161.1.31

Pfister, S., Koehler, A., & Hellweg, S. (2009). Assessing the environmental impacts of freshwater consumption in LCA. *Environmental Science and Technology, 43*, 4098–4104.

Raworth, K. (2013). Defining a safe and just space for humanity. In *Is sustainability still possible? State of the World 2013*. Washington DC: The Worldwatch Institute.

Raworth, K. (2012). A safe and just space for humanity: Can we live within the doughnut? *Oxfam Policy and Practice: Climate Change and Resilience, 8*, 1–26. doi:10.5822/978-1-61091-458-1

Recover. (2016). Recover Products [WWW Document]. http://www.recovertex.com/products. Accessed 1 March 2016.

Rees, W., & Wackernagel, M. (1996). Urban ecological footprints: Why cities cannot be sustainable—and why they are a key to sustainability. *Environmental Impact Assessment Review, 16*, 223–248.

Robinson, J. B. (1982). Energy back-casting. A proposed method of policy analysis. *Energy Policy, 10*, 337–344.

Roos, S., Posner, S., Jönsson, C., & Peters, G. M. (2015a). Is unbleached cotton better than bleached? Exploring the limits of life cycle assessment in the textile sector. *Clothing and Textiles Research Journal*. doi:10.1177/0887302X15576404

Roos, S., Sandin, G., Zamani, B., & Peters, G. M. (2015b). Environmental assessment of Swedish fashion consumption. Five garments—sustainable futures. Mistra Future Fashion, Stockholm, Sweden.

Roos, S., Zamani, B., Sandin, G., Peters, G. M., & Svanström, M. (2016). An LCA-based approach to guiding an industry sector towards sustainability: The case of the Swedish apparel sector. *Submitted to Journal of Cleaner Production.*

SAC. (2016). Sustainable Apparel Coalition (SAC) [WWW Document]. http://apparelcoalition. org/the-higg-index/. Accessed 24 August 2013.

Sandén, B. (2012). Standing the test of time: Signals and noise from environmental assessments of energy technologies. Based on a paper presented at the Symposium on Life Cycle Analysis for New Energy Conversion and Storage Systems, Materials Research Society Symposium Proceedings (Vol. 1041). Boston, MA, 2008.

Sandén, B. A., & Hedenus, F. (2014). Assessing biorefineries. In *Systems Perspectives on Biorefineries* (pp. 6–17). Gothenburg, Sweden.

Sandin, G., Peters, G. M., & Svanström, M. (2015). Using the planetary boundaries framework for setting impact-reduction targets in LCA contexts. *International Journal of Life Cycle Assessment, 20,* 1684–1700. doi:10.1007/s11367-015-0984-6

Sandin, G., Peters, G. M., & Svanström, M. (2013). Moving down the cause-effect chain of water and land use impacts: An LCA case study of textile fibres. *Resources, Conservation and Recycling, 73,* 104–113.

Schellenberger, S., Holmquist, H. M., Berger, U., Cousins, I. T., Gillgard, P., Leonards, P., et al. (2015). Substitution of long chain fluorinated copolymers for durable water repellent (DWR) textile modification, In *Society of Environmental Toxicology and Chemistry (SETAC)| SETAC Europe, 25th Annual Meeting,* Barcelona.

Schenk, R. (2013). Sustainable Apparel Coalition Product Category Rule Guidance 2013.

Shen, L., Worrell, E., & Patel, M. K. (2010). Environmental impact assessment of man-made cellulose fibres. *Resources, Conservation and Recycling, 55,* 260–274. doi:10.1016/j. resconrec.2010.10.001

Stamminger, R. (2005). Preparatory studies for Ecodesign requirements of EuPs. Lot 14: Domestic washing machines and Dishwashers. Bonn, Germany.

Statistics Sweden. (2014). Statistics database [WWW Document]. http://www.scb.se/sv_/Hitta-statistik/Statistikdatabasen/

Steffen, W., Richardson, K., Rockström, J., Cornell, S., Fetzer, I., Bennett, E., et al. (2015). Planetary Boundaries: Guiding human development on a changing planet. *Science* (80), 347. doi:10.1126/science.1259855

Swedish EPA (2015). Sweden's Environmental Objectives [WWW Document]. http://www. miljomal.se/Environmental-Objectives-Portal/. Accessed 8 October 2015.

UNEP. (2009). Guidelines for social life cycle assessment of products. New York and Geneva.

UNFCCC. (2016). United Nations Framework Convention on Climate Change [WWW Document]. http://unfccc.int/essential_background/items/6031.php. Accessed 30 March 2016.

United Nations. (2016). Montreal Protocol on Substances that Deplete the Ozone Layer [WWW Document]. http://ozone.unep.org/en/treaties-and-decisions/montreal-protocol-substances-deplete-ozone-layer. Accessed 1 March 2016.

United Nations. (2015a). World population prospects, the 2015 revision [WWW Document]. http://esa.un.org/wpp. Accessed 1 March 2016.

United Nations. (2015b). Sustainable Development Goals [WWW Document]. http://www.un.org/ sustainabledevelopment/poverty/. Accessed 30 October 2015.

Wackernagel, M., Onisto, L., Bello, P., Linares, A. C., López Falfán, I. S., Méndez García, J., et al. (1999). National natural capital accounting with the ecological footrpint concept. *Ecological Economics, 29,* 375–390.

Advisors, Wazir. (2015). *Strategy for enhancing India's share in global textile and apparel trade.* India: Haryana.

World Commission on Environment and Development. (1987). Our Common Future.

Zamani, B. (2014). Towards understanding sustainable textile waste management: Environmental impacts and social indicators. Chalmers University of Technology.

Zamani, B., Sandin, G., & Peters, G. M. (2016a). Life cycle assessment of clothing libraries: Can collaborative consumption reduce the environmental impact of fast fashion? *Submitted to Journal of Cleaner Production.*

Zamani, B., Sandin, G., Svanström, M., & Peters, G. M. (2016b). Hotspot identification in the clothing industry using social life cycle assessment—opportunities and challenges of input-output modelling. *The International Journal of Life Cycle Assessment* (In print).

ZDHC. (2014). Roadmap to Zero Discharge of Hazardous Compounds (ZDHC) [WWW Document]. http://www.roadmaptozero.com/. Accessed 1 June 2014.

Sustainability in the Textile and Fashion Industries: Animal Ethics and Welfare

Miguel Ángel Gardetti

Abstract There is no doubt that the textile (and fashion) industry is important for the economy; however, taking into account the concept of sustainability, this industry many times—actually most times—operates to the detriment of environmental and social factors. John R. Ehrenfeld defines sustainability as *"the possibility that humans and other life will flourish on the earth forever"* (Ehrenfeld and Hoffman 2013, p. 7). This notion of sustainability is not only a concern for people and the environment, but also for animals. Besides environmental and social issues, more than 50 million animals suffer cruel death each year to benefit the fashion industry (Born Free USA 2014). The purpose of this chapter is to make a contribution to the animal care agenda in the textile and fashion industry by presenting the United Nations Global Compact Code of Conduct for the Textile and Fashion Industry and Ovis 21 case—a company from the Argentine Patagonia, that is a B Corp and Savory Institute Hub company, but a company in which People for the Ethical Treatment of Animals (PETA) discovered cruelty to animals in one of their establishments. This also puts forward the reactions of two of its most important clients, the media and consumers. This chapter closes with an analysis and some conclusions about the topic and Ovis 21 case.

Keywords Sustainability · Animal ethics and welfare · UN Global Compact · Code of Conduct and Manual for the Fashion and Textile Industry: the First "Sectorial" Initiative of the Global Compact · Ovis 21

M.Á. Gardetti (✉)
The Center for Studies on Sustainable Luxury, Av. San Isidro 4166, Ground Floor "A",
C1429ADP Buenos Aires, Argentina
e-mail: mag@lujosustentable.org
URL: http://www.lujosustentable.org

© Springer Science+Business Media Singapore 2017   47
S.S. Muthu (ed.), *Textiles and Clothing Sustainability*, Textile Science
and Clothing Technology, DOI 10.1007/978-981-10-2182-4_2

1 Introduction

Because of the size of the sector and the historical dependence of clothing manufacture on cheap labor, the clothing and textile industry is subject to intense political scrutiny and has been significantly shaped by international trade agreements.

Estimating the number of people that work in these sectors is extremely difficult because of the number of small firms and subcontractors active in the area, as well as the difficulty in drawing boundaries between sectors (Allwood et al. 2006). According to statistics from the United Nationals Industrial Development Organization (UNIDO) Industrial Statistics Database (INDSTAT), around 26.5 million people work in the clothing and textiles sector worldwide (International Labour Organization–ILO 2006).

The above-mentioned ILO report further states that of these 26.5 million employees, 13 million are employed in the clothing sector and 13.5 million in the textile sector. These figures are only people employed in manufacturing—not retail or other supporting sectors (Allwood et al. 2006).

Against this background, there is no doubt that the textile (and fashion) industry is significant for the economy; however, taking into account the concept of sustainability, this industry many times—actually most times—operates to the detriment of environmental and social factors (Gardetti and Torres 2011). Moreover, over the past 25 years, a slow but radical change has taken place within the fashion world toward more sustainable production.

John R. Ehrenfeld defines sustainability as *"the possibility that humans and other life will flourish on the earth forever"* (Ehrenfeld and Hoffman 2013, p. 7). And he explains, *"I took flourishing as a metaphor that captures happiness, health, and many characteristics of what humans believe is a good life. And it captures a sense of the health of the natural world"* (Ehrenfeld and Hoffman 2013, p. 22).

This notion of sustainability is a concern not only for people, but also for animals and the environment. And, along this line, the International Standard ISO 26000 (2010, p. 11–13) considers the welfare of animals, including the provision of decent conditions for keeping, breeding, producing, transporting, and using animals.

While those who have been drawing attention to animal welfare issues have been ridiculed as emotional activists and extremists, the purpose of this chapter is to make a contribution to the animal care agenda in the textile and fashion industry by presenting the United Nations Global Compact Code of Conduct for the Textile and Fashion Industry and Ovis 21 case—a company from the Argentine Patagonia, that is a B Corp and Savory Institute Hub company—engaged in the regeneration of grasslands and specialized in holistic management, with broad experience in sheep and wool. People for the Ethical Treatment of Animals (PETA) organization discovered cruelty to animals in one of Ovis 21 establishments. It is worth mentioning that this chapter is not intended to delve into animal rights.

2 Methodology

To develop this chapter, the author based on the UN Global Compact's Code of Conduct and Manual for the Fashion and Textile Industry and, particularly, on Principle 11 on animal care. For Ovis 21 case development and analysis, he collected background information from the company and PETA organization. Regarding System B—an organization that certifies the overall social and environmental performance, legal accountability and public transparency—and Savory Institute—an organization that facilitates the realization of a life of enduring returns for the land and all who depend on it—the author exchanged information via e-mail and also resorted to mass media, such as The Huffington Post (UK), Women's Wear Daily (WWD)—a property of Fairchild Fashion Media, EcoTextiles Magazine (UK), and the Herald Premium (Spain).

3 Sustainability and Animal Ethics and Welfare in the Textile and Fashion Industries

Several authors and organizations have analyzed the impacts of the textile and clothing industry. Some of them are Slater (2000), Allwood et al. (2006), Fletcher (2008), the UK Department for Environment, Food and Rural Affairs (2008), Ross (2009), Dickson et al. (2009), and Gwilt and Rissanen (2011).

One specific study has been Fashioning Sustainability—A review of sustainability impacts of the clothing industry, which Stephanie Draper, Vicky Murray, and Ilka Weissbrod conducted in 2007 for the World Wild Fund, financed by Mark & Spencer.

Along this line, we bring the attention, for instance, to how fiber as a raw material is obtained, the use of pesticides during this process that leads to health problems for workers, causes soil degradation and biodiversity loss. Also attention is drawn to water which is such a vital element in the processing of cotton, in particular, so much so that this crop has been the so-called thirsty crop. In turn, while the use of agrochemicals tends to be reduced, the use of genetically modified organisms could lead to another type of impacts.

Abuses on working conditions are also commonly presented in other stages of these industries; many times, human rights are violated in the so-called sweatshops which are characterized by low wages and long hours. The risks are even greater if safety, security and healthcare systems are not appropriate.

In turn, many of the synthetic fibers are derived from non-renewable resources such as oil. In general, environmental abuse combines with ethical issues when there is an overuse of water and when land for food production is taken over.

Considering the whole textile chain—from spinning all the way through to finishing—it cannot be ignored that the use of chemicals may have carcinogenic and neurological effects, may cause allergies, and may affect fertility. During these processes, large amounts of water and energy are used, and in general, non-biodegradable wastes are produced.

In the marketing and sales processes, there are subsidies and quotas with huge impact on developing countries. Moreover, the lack of international regulation on these issues creates a win/lose scenario, and prices should allow for fair profit-sharing throughout the supply chain. All of these stages also involve the use of energy and lots of packaging, as well as the generation of carbon emissions (CO_2). The paradox, in this case, is that for survival, the working force depends on a system that seems to be destroying the world's capacity to withstand it. In both textile and fashion design, sustainability is, in general, perceived as an obstacle.

And last, but not least, major impacts arise from transport, such as carbon emissions and waste generation.

Environmental and social issues are also present with more than 50 million animals suffering cruel deaths each year to benefit the fashion industry (Born Free USA 2014). The fur industry on its own currently kills around 30 million animals a year (Animal Equality, s.d.). These numbers are really high because neither fur nor fur trim is a by-product of the meat industry; fur comes from animals that are factory-farmed or trapped purely for fashion (Born Free USA 2014). Animals on those fur farms live in cramped, dirty cages and are killed using the cheapest methods, such as suffocation, gassing, or electrocution. Another scandal is that of leather coming from countries, such as India and China, where the animal throats are cut and their skin ripped off while they are still alive. This is far from the definition of animal welfare coined by Hughes (1988, p. 33), as a state of physical and mental health where an animal is completely in harmony with its environmental surroundings.

According to Dobson (2003), consumers need to ponder about the implications of their routine purchasing decisions and change behaviors. This was called "ecological citizenship" by Dobson (2003). However, because most organizations are not transparent, consumers are not always aware of the damage they cause to animals and the environment. Combining sustainability and fashion is a true challenge (Molderez and De Landtsheer 2015).[1] This attitude is important since Teresa Presas in her work "Interdependence and Partnership: Building Blocks to Sustainable Development" (2001) exposes that a real transition toward sustainable development requires a new way of thinking. It requires the use of a collective learning mechanism for all types of environments and stakeholders, and structured discussions about our vision of what a sustainable society is all about. But a sustainable society is not possible without sustainable individuals (Cavagnaro and Curiel 2012). That is, individual capacities seem to be at the heart of this issue. These definitions should lead to a more informed and responsible attitude from consumers.

---

[1]According to Moderez and De Landtsheer (2015), the Chinese fur is often intentionally mislabeled, so when the customers buy fur, they are unable to trace where it comes from.

3.1 The UN Global Compact

The UNGC is the result of a world characterized by glaring and unsustainable imbalances and inequities (Kell 2003). This is a joint initiative of the United Nations Development Program, the Economic Commission for Latin America and the Caribbean, and the World Labor Organization, in an effort to enable corporate social responsibility development and to foster human rights, labor standards, environmental protection, and anticorruption. The main goal of the UNGC is to help align corporate policies and practices to universally concurred and internationally applicable ethical goals, and it is based on principles that stem from four key agreements: the Universal Declaration of Human Rights, the International Labour Organization's Declaration on Fundamentals Principles and Rights at Work, the Rio Declaration on Environment and Development, and the United Nations Convention Against Corruption. That is to say, by means of business voluntary commitment, the Global Compact is an initiative for promoting a new corporate culture on how to manage businesses. Its real essence is to create an ever-growing labor network (McIntosh et al. 2004a, b) supporting companies through learning and knowledge sharing, exercising leadership as a corporate citizen, and hence exerting influence on others through their behavior (Fuertes and Goyburu 2004).

3.1.1 Code of Conduct and Manual for the Fashion and Textile Industry: The First "Sectorial" Initiative of the Global Compact

On May 3, 2012, the United Nations Global Compact presented its first sector-specific initiative. This is a Code of Conduct and Manual for the Fashion and Textile Industry jointly developed with the Nordic Fashion Association and the Nordic Initiative Clean and Ethical—NICE. This presentation took place at the Copenhagen Fashion Summit of that year, and George Kell—Executive Director of the UN Global Compact—said: "*As an industry facing serious and widely publicized social and environmental challenges, the fashion and textile industry is uniquely positioned to launch a sectorial initiative under the umbrella of the UN Global Compact. We are very excited about this effort and look forward to collaborating with NICE and its partners*" (Nordic Fashion Association 2012).

There was a second launch during the Rio+20 Summit with the purpose of boosting this new initiative in the activity titled "Changing the World through Fashion: Contribution to Sustainable Development by the Fashion and Textile Industry." In this new activity, Mr. G. Kell emphasized "*the importance of changing the fashion and textile sector*," and said that "*this new impetus was backed by the Global Compact.*"

Why a Code of Conduct? Mina Kaway in her paper prepared in 2009 titled "Corporate Social Responsibility through Codes of Conduct in the Textile and Clothing Sector" states (p. 7): "*not only due to the subcontracting chain system used in this industry but also due to the fact that the Textile and Clothing industry is mainly a labor industry, that is, the use of manual workers is high if compared to other machinery manufacturers sectors, and therefore creating an environment more favorable for abuses to occur in this sense."*

While this UNGC Code includes the ten principles of the United Nations Global Compact, it provides additional specificity from a sector perspective, adding 6 principles within a varied range of topics that pertain to the fashion and textile industry relative to an area called ethical conduct. Table 1 shows the 16 principles of the code, matching them with the compact, sector specificity, and relevant areas.

Table 1 UNGC code of conduct: Principles and subject areas

Code of Conduct principles		
1. Support and observe human right protection	Global Compact	Human rights
2. Not to be an accomplice to abuse of rights		
3. Support the principles of freedom of unionization and the right to collective bargaining		Labor rights
4. Eradicate forced and obligatory labor		
5. Abolish any form of child labor		
6. Eliminate discrimination based on job and occupation		
7. Support a preventive approach to environmental challenges		Environment
8. Foster greater environmental responsibility		
9. Promote development and dissemination of green technologies		
10. Businesses must act against corruption in all its forms, including extortion and bribery		Anti-corruption
11. Animals must be treated with dignity and respect. No animal must be deliberately harmed or exposed to pain	Sector Specificity	Ethical conduct
12. Businesses and their designers must work actively to encourage and support sustainable design		
13. Businesses must, through their choice and treatment of models, promote a healthy lifestyle and healthy body ideals, and the models' minimum age must be 16 during fashion weeks and other occasions where the workload is excessive		
14. Businesses must work toward transparency in their supply chain		
15. Businesses must work toward a stronger commitment throughout their supply chain to reinforce the development of a secure mining industry		
16. All businesses involved must at all times be open and accessible for announced, semiannounced, and unannounced audits for monitoring and evaluation of compliance with the Code of Conduct		

Source Prepared by the author
For each of these principles, the NICE Code of Conduct and Manual for the Fashion and Textile Industry at the Nordic Fashion Association; Nordic Initiative Clean and Ethical—NICE—and Global Compact (2012): explains **What** it means to act in accordance with the Code of Conduct (i.e., the "goal" of each principle) **Why** each principle is important **Learn more**, and **How** (recommendations)

Furthermore, you will find important facts and information on where to learn more about creating a long-term viable and sustainable business. Note that the code is applicable not only to the brand (or company owner of the brand), but also to every partner, both in the country and in abroad. According to the code, the brand is responsible for developing an ethical and sustainable supply chain in the company.

Likewise, the code includes some peculiarities, which are described below:

(a) In the *labor right* area, it delves deeper into *working hours; employment contracts; sick leave and paid vacations; complaint filing system; occupational safety and health;* and *home work.*
(b) In the *environment* area, and in connection with principles 7 and 8, it highlights the *waste and water* topic, while referring to—within the framework of principle 9—*the use of chemicals, energy, carbon dioxide emissions, and atmospheric emissions in general.*
(c) As to *monitoring and evaluation*, it provides an analysis of the value chain development based on three levels of risk—basic, high, and advanced, suggesting some guidelines for each of them, namely:

Basic Level

- Include CSR clauses or the UNGC Code of Conduct in supplier contracts and begin working toward integration.
- Conduct informal (code-based) audits when visiting suppliers for other reasons.

High Level

- Make a detailed mapping of all suppliers so as to assess them in accordance with their specific risk level. Operating in the fashion industry typically means maneuvering in high-risk countries.[2]
- When your supply chain management program has been established, you can use the following generic risk matrix to determine whether it will be enough for the supplier to sign the NICE Code of Conduct, or whether you will also need to audit suppliers regularly. When designing the audit program, you will need to determine how much commitment and willingness to work with different suppliers your company has.

Advanced Level

- Send out a self-evaluation questionnaire, partly to get an initial knowledge of the suppliers' performance level within CSR and partly to point out the requirements in the Code of Conduct.
- Create good dialog with suppliers so that they perceive the self-evaluation process as part of their long-term relationship.

---

[2]Here, the UNGC Code of Conduct suggests considering the 'Danish Institute for Human Rights' to use Human Rights and Business country risk analysis.

- Conduct formal audits solely concentrating on environmental, social, and ethical issues. Both announced or unannounced audits are possible, each of which has various advantages.
- It states that *risks* may be influenced by various factors, including spending, country, category, and the transaction nature, in addition to how critical the supplier is to your business. Broadly speaking, the more critical the supplier is, the higher the overall risk will be. Therefore, suppliers need to be divided into three categories, depending on their criticality:
- highly critical (it means that replacing the supplier would be extremely costly and disruptive),
- semi-critical (it means that replacing the supplier is possible, but it is time-consuming and partly costly),
- less critical (suppliers can be replaced on a need basis,)
- It suggests that an *audit* should begin with a meeting with the supplier's management (including the person responsible for implementing the code, an HR representative, and even the local union representative) where the audit outline is reviewed and discussed. This meeting should deal with the "how's" explained in each principle within the framework of the Code.

4 Code of Conduct and Manual for the Textile and Fashion Industry's Principle 11: Animal Ethics and Welfare[3]

Principle 11| Many fashion businesses have made a conscious ethical decision not to use real animal fur (e.g., fax, sable, mink, and rabbit) or exotic and wild-caught animal species (e.g., snake, crocodile, and ostrich). We recognize this choice, and we acknowledge that other businesses have chosen to take a different path. In businesses where animals are used in labor and/or in the production (fur, wool, etc.), such animals must be fed and treated with dignity and respect and no animal must deliberately be harmed nor exposed to pain in their life span. Taking the lives of animals must at all times be conducted using the quickest and the least painful and non-traumatic (not in the vision of other animals) method available and approved by national and acknowledged veterinarians and only conducted by trained personnel.

What
We do not tolerate the maltreatment of animals, and animals must be cared for and protected from harm. We do not support the use of any endangered species listed in

---

[3]Reproduction of Principle 11 contents NICE Code of Conduct and Manual for the Fashion and Textile Industry at the Nordic Fashion Association; Nordic Initiative Clean and Ethical—NICE—and Global Compact (2012, p. 69).

the International Union of Conservation of Nature (IUCN) Red List of Threatened Species. We recommend following the guidelines in the European Convention for the protection of animals kept for farming purposes.

Why

Animals are sentient beings, and it is the responsibility of humans to ensure that they have a "life worth living." We do not support the usage of down and feather plucked from living birds. The maltreatment of animals can cause severe reputational damages in relation to retailers, consumers, and other stakeholders. Animal activists are very persistent in their work and have a record of great influence on decision-makers.

Learn more

European regulation on the protection of farming animals, animal welfare, and industry regulations: The EC strategy 2012–2015.

Information on threatened animals: IUCN Red List of Threatened Species.

International guidelines on international governmental recommendations and regulative instruments: UN Food and Agricultural Organization (FAO).

How

Have an animal treatment policy that clearly states that garments containing animal-derived products are produced using abundant species that have been treated in accordance with international animal welfare standards, as well as animal welfare standards laid down by Europe law.

Clearly label garments containing parts of animal origin as such, including the name of the part used (such as leather or natural fur) to ensure that consumers are not deliberately or unintentionally mis-sold goods they do not wish to purchase.

Species farmed for any consumer goods must be produced to standards found on highly keeping of regulated European farms. This included Directive 98/58 on the protection of animals kept for farming and the 1999 Council of Europe Recommendations on the keeping of animals for fur. Animals taken from the wild must have been afforded the protection of the International Agreement on Humane Trapping Standards and hunted in accordance with the International Union Conservation of Nature's "sustainable use" policy.

Where possible, reputable voluntary schemes should be used to ensure that the highest possible standard of care is given to all animals used for the purposes of fashion.

Always obtain a guarantee that down and feather only originate solely from dead birds.

This principle has its corresponding NICE code shown in Exhibit 1.[4]

Both producers and suppliers should consider the fact that wild animals and their habitats are part of their natural ecosystems and should therefore be valued and

---

[4]The contents of Principle 11, the NICE Code of Conduct Principles—Nordic Fashion Association and Nordic Initiative Clean and Ethical—NICE—and Global Compact (2012, p. 99) are reproduced.

protected and that their welfare must be taken into account (International Standard ISO 26000, 2010, p. 47).

5   Sustainability with Cruelty to Animals? The Case of Ovis 21 (Argentina)

5.1   Ovis 21: The Company[5]

Ovis 21 is a B Corp engaged in regenerating grasslands and increasing business profitability. It provides training, consulting, and product certification services. As a Hub of the Savory Institute, it specializes in holistic management and has great experience in sheep and wool. It has developed and manages a network of innovative land managers, technicians, and related industries to improve sustainability aspects.

History, Mission, and Philosophy
Ovis 21 was created in the year 2003 by Pablo Borrelli, Ricardo Fenton, and Estancia Monte Dinero to improve sustainability in sheep-based value chains. Pablo Borrelli with his wife Alejandra Canosa and Ricardo Fenton founded an organization which would be capable of introducing innovations at a regional level, making use of the capabilities of entrepreneurial farmers and professionals in the region.

The stewardship and improvement of natural grasslands was one of the ongoing priorities of the network founders. The introduction of "holistic management" in 2007 has enabled to work on grassland regeneration at full farm scale.

In collaboration with the Nature Conservancy, Ovis 21 developed the Grassland Regeneration and Sustainable Standard (GRASS), as well as flock improvement and advanced wool classing protocols. Ovis 21 is responsible for certifying farmers' compliance with these protocols. There are over 3.2 million acres GRASS certified (1.3 million hectares) in 54 different properties.

They operate through independent representatives who are accredited by the company to provide advice and coordinate the provision of services under the protocols developed by Ovis 21.

Today, the network has more than 160 primary producers and 22 studs distributed in the Argentine provinces of Santa Cruz, Tierra del Fuego, Chubut, Río Negro, Neuquén, Buenos Aires, and Corrientes, as well as in the south of Chile and Uruguay.

Ovis 21 is part of the 15 million Acre Campaign in collaboration with Patagonia Inc. and the Nature Conservancy, aiming at the regeneration of temperate grasslands in the Patagonian region.

---

[5]Based on Ovis 21 www.ovis21.com Accessed: 1 Mar 2016.

Its mission is to promote a culture of grassland biodiversity regeneration, so that land will be capable of sustaining people, their businesses, and communities. To achieve this, Ovis 21 focuses on promoting decision-making to regenerate land, increase biodiversity, develop a land stewardship culture, empower people to fully develop their capacities and businesses, and enhance collaboration as the means to meet goals that cannot be otherwise attained individually.

Part of Ovis 21 philosophy includes integrating the satisfaction of real needs, the creation and distribution of wealth, the development of people and organizations, the collaboration based on trust, a holistic view, and an entrepreneurial spirit as a source of continuous innovation for sustainability.

5.2 Two Key Organizations Within the Work Framework of Ovis 21[6]

B Corp. Individually, they meet the highest standards of verified social and environmental performance, public transparency, and legal accountability and aspire to use the power of markets to solve social and environmental problems. Collectively, B Corp lead a growing global movement of *people using business as a force for good^{TM}*. Through the power of their collective voice, one day all the companies will compete to be *best for the world^{TM}*, and society will enjoy a more shared and durable prosperity for all. Note that part of Ovis 21 declaration of interdependence says: "*through their products, practices, and profits businesses should aspire to do no harm and benefit all.*"

The Savory Institute facilitates the realization of a life of enduring returns for the land and all who depend on it. The institute is the brain trust of the organization. It develops innovative tools and enhanced curricula, informs policy, establishes market incentives, increases public awareness, and coordinates relevant research, cultivating relationships with aligned partners.

Their strategy is threefold: to demonstrate results in diverse contexts; to equip and empower people and share their experiences in learning and achieving success; and to pave the way for a global movement to take off, being an advocate and catalyst of needed change.

5.3 Ovis 21 and the Animal Ethics and Welfare: An Episode

According to Gordon and Hill (2015), People for the Ethical Treatment of Animals (PETA) organization which was founded in 1980 and is well known for its

---

[6]Based on B Corp Web site (https://www.bcorporation.net/) and the Savory Institute's Web site (http://savory.global/) Accessed: 1 Mar 2016.

provocative activism has been one of the most outspoken champions of animal rights over the past three decades. According to Paulins and Hillery 2009, this organization believes that animals have rights and should not be abused for the sake of fashion.

PETA[7] focuses its attention on the four areas in which the largest number of animals suffers the most intensely for the longest periods of time: on factory farms, in the clothing trade, in laboratories, and in the entertainment industry. We also work on a variety of other issues, including cruel killing of rodents, birds, and other "pests" as well as cruelty to domesticated animals. The organization also works through public education, cruelty investigations, research, animal rescue, legislation, special events, celebrity involvement, and protest campaigns.

On August 13, 2015, PETA uploaded on "youtube.com" a video which shows the cruelty taking place in a farm called Estancia La Librun, that is part of the Ovis 21 network in Argentina and which provides wool to several renowned brands, such as Patagonia and Stella McCartney. On August 13, PETA sent a formal letter to the Office of Wild Fauna and Protected Areas of Argentina signed by Elisabeth Custalow, lawyer of PETA Foundation. The letter—along with video-recorded and photographed evidence—requested that research be carried out on the aforementioned *Estancia* and, if needed, that the pertaining charges be pressed against those in charge.[8] Peta director, Mimi Bekhechi, said: "*Having witnessed an astounding level of cruelty to sheep in every shearing shed Peta US visited Australia, the US and, now, Argentina, it's high time that companies and consumers ditch real wool. Today, finding alternatives to hide, fur and fleece is easy, and no animal has to suffer when businesses make kind choices* (Snowdon 2012)."

Due to PETA's position,[9] the episode became notorious on the media. Some international examples are The Huffington Post (UK), Women's Wear Daily (WWD) —a property of Fairchild Fashion Media, EcoTextiles Magazine (UK), and the Herald Premium (Spain), the Quartz Daily Brief, Women's Wear Daily, and Fashionista.

5.4 The Decision of Various Important Clients

On "August 12, 2015"—a day before PETA uploaded its video on youtube.com, Tessa Byars[10] from Patagonia presented a detailed history of the company's work

---

[7]People for the Ethical Treatment of Animals—PETA Source: PETA's Web site www.peta.org Accessed 1 Mar 2016.

[8]Exchange of e-mails with Ms. **Hannah Schein**, associate director of Cruelty Investigations| People for the Ethical Treatment of Animals (PETA). The author has a letter presented by PETA.

[9]According to Paulins and Hillery (2009) before 2009, they had a dispute with the Australian wool industry for the practice of "mulesing" including boycotts against Abercrombie & Fitch, J. Crew and Benetton. To promote this campaign, PETA also recruited celebrities, such as singer Pink and Australian actress Toni Colette.

[10]See Patagonia http://www.patagoniaworks.com/press/2015/8/12/petas-wool-video Accessed: 1 Sept 2016.

with Ovis 21. Previously, she highlighted her conviction about the company's defense and good treatment of animals. However, this "statement"—which can be seen in full length in Exhibit 2[11]—said, *"We will work with Ovis 21 to make needed corrections and improvements, and report back to our customers and the public on the steps we will take."* Concluding, she added, *"We apologize for the harm done in our name; we will keep you posted."*

A few days after, on August 17, 2015, Rose Marcario, the company's CEO, said in a statement, *"We took some important steps to protect animals in partnering with Ovis 21, but we failed to implement a comprehensive process to assure animal welfare, and we are dismayed to witness such horrifying mistreatment."* She also made public the company's decision to discontinue purchases from Ovis 21 and explained the reason *"...we've made a frank and open-eyed assessment of the Ovis program. Our conclusion: it is impossible to ensure immediate changes to objectionable practices on Ovis 21 ranches."* For the full statement, see Exhibit 3.

Indeed, Stella McCartney has suspended all purchases of wool from Ovis 21 after watching the video. With a statement released through PETA (qz.com. 2015), McCartney said on August 13 (PETA 2015):

> After conducting our own investigation in Argentina and throughout our supply chain, following a very distressful viewing of PETA US footage, we immediately ceased buying wool from Ovis 21. We are deeply saddened and shocked by the cruelty seen on the footage, as animal welfare is at the heart of everything we do. This is a huge setback to help saving the grasslands in Patagonia. We are now even more determined to continue our fight for animal rights in fashion together and monitor even more closely all the suppliers involved in this industry. We are also looking into vegan wool as well, in the same manner we were able to develop and incorporate high-end alternatives to leather and fur over the years.

In turn, on Instagram, Stella McCartney added, *"Unfortunately, after conducting our own investigation in Argentina, following a very distressful viewing of footage provided by the great guys at @officialpeta, we found out that 1 of the 26 ranches we used to source sustainable wool there, mistreated its sheep. It is one too many."* Exhibit 4 shows two posts from McCartney on Instagram.

Moreover, Peta director—Mimi Bekhechi—said, *"Having witnessed an astounding level of cruelty to sheep in every shearing shed Peta US visited Australia, the US and, now, Argentina, it's high time that companies and consumers ditch real wool. Today, finding alternatives to hide, fur and fleece is easy, and no animal has to suffer when businesses make kind choices* (Snowdon 2012). *We thank Stella for rejecting cruelty in the wool industry and hope Patagonia and others will follow suit"* (Snowdon 2012).

Kering Group also agreed to stop sourcing from Ovis 21 for all brands, including Alexander McQueen, Balenciaga, Bottega Veneta, Brioni, Christopher Kane, Gucci, Volcom, and Saint Laurent.

---

[11]Source http://www.thecleanestline.com/2015/08/patagonia-to-cease-purchasing-wool-from-ovis-21.html.

5.5   The Ovis 21 Statement[12]

On August 15, 2015, Ovis 21 posted the unsigned statement shown below on its Web site (see screenshot on Exhibit 5):

> Ovis 21 certifies grassland regeneration, flock improvement and wool quality in a network of farms.

> With regard to the video footage taken in December 2014 in one of the farms in the Ovis 21 network, the images depicting inhumane treatment of lambs and sheep are unacceptable. Ovis 21 does not justify cruelty. We have identified and intervened the property involved, which is now no longer a certified property. We regret not being informed when the footage was taken, to take immediate action.

> Farming practices have a relevant role as a tool to improve biodiversity and ecosystem function, to produce food and fiber for an increasing world population, and to generate the best environment for wildlife, rural families, and their communities.

5.6   The Standing of the Two Key Organizations After the Episode

The author exchanged mails with two organizations with which Ovis 21 had a close relationship, the Savory Institute and System B (and B Lab). Both knew that the author was preparing this chapter and that it comprised the analysis of Ovis 21 case. The former organization did not allow the use of information collected during said exchanges and referred to its Web site since it is their official site.[13] System B had— along with the author—a different approach: open and it allowed for the use of the information exchanged.[14]

The author intended to check whether they had (or may have) made some kind of public statement since evidence provided by PETA was totally irrefutable for two reasons:

- First, that Patagonia (a B Corp) and Stella McCartney had made their own audits and proved the existence of "cruelty."
- And second, Ovis 21 own statement posted on their Web site, which fully accepted the facts (that is to say that there was, indeed, cruelty to animals in one of their establishments).

[12]Source: Ovis 21 Web site.
[13]e-mail sent from contact@savory.global November 11, 2015. e-mails were always signed "by the institution" under the name The Savory Institute Team.
[14]e-mail from Christina Foorwod—Senior Associate, Standards—November 17, 2015.

At the time of writing this chapter, none of the organizations issued a public statement. Moreover, System B, which gave a detailed account of the process, declared that[15]: "...*We will be happy to provide more information as soon as we have completed our process of reviewing Ovis's recertification.*"

The screenshot on Exhibit 5 shows that the logos of both institutions are still on Ovis 21 Web site as of March 10, 2016.

6 Sustainability and Animal Ethics and Welfare: Analysis and Conclusions

The United Nations Global Compact has been working thoroughly to promote a change in the current textile and fashion system. An example of this is the Code of Conduct and Manual for the Fashion and Textile Industry. This piece clearly shows that in fashion businesses, animals must be fed and treated with dignity and respect, and no animal must deliberately be harmed nor exposed to pain in their life span.

In business, value creation is quickly advancing from the so-called tangible assets to intangible assets, such as intellectual, social, and human capital. However, the relationship between business and stakeholders is an important but not yet valued corporate asset. Technology and globalization are creating working networks which are decisive corporate assets. Kevin Kelly (1999) reinforces this view when he argues in his paper "New Rules for the New Economy" that network economy is supported by technology, but that it can only be built on the basis of "relationships." It begins with "chips" and it ends with "trust" building.

It is worth noting that loyalty toward a brand has been recognized as an intangible asset. An increased awareness of the relationship between intangible assets, such as customer satisfaction, customer loyalty, and brand loyalty, has made it easier to estimate the financial value of brands. These are some of the reasons for Stella McCartney and Patagonia's reactions, even though they have been different. These concepts emerged more quickly in McCartney and are shown in two posts on Instagram: "*We found out that 1 of the 26 ranches we used source sustainable wool there, mistreated its sheep. It is one too many*" and "*As a designer who built a brand on not using leather, fur or animal skins in its designs, I can't tolerate it!*"

Led by Patagonia—when it became aware of PETA campaign about Australian wool in 2005—and along with Ovis 21, both organizations might have come closer to the so-called unconventional stakeholders, including, among others, the organizations that defend animal rights, which may be "divergent" as their interests are against those of the company. For both organizations, this would entail to gain the ability to reconcile contradictions between the current business model and the opinions of "unconventional" stakeholders (Hart and Sharma 2004). Approaching PETA would help Patagonia and Ovis 21 access knowledge and

---

[15]e-mail from Christina Foorwod dated October 19, 2015.

prospects which are essential both to anticipate potential sources of problems and to identify innovative opportunities and business models for the future (Hart and Sharma 2004).[16]

The main reason presented by Patagonia to cancel purchases from Ovis 21—according to its CEO's statement—was that "...*we've made a frank and open-eyed assessment of the Ovis program. Our conclusion: it is impossible to ensure immediate changes to objectionable practices on Ovis 21 ranches.*" Yet, the pressure exerted by consumers in general, animal advocates, and Patagonia's customers was so great that it jeopardized its reputation when it stated that its "suppliers respect welfare." Exhibit 6 shows statements from different sources.

The stances of the Savory Institute and System B clearly affect their image, reputation, transparency, and effectiveness. The formal "absence" of these organizations in the face of Ovis 21 case may most likely be interpreted by many other organizations and individuals as lack of transparency, sending a totally wrong message to the society.

The development within the new paradigm of the sustainable mind-set in the fashion industry makes it inevitable not to counteract ethics and morality. Actually, every choice made in the business of fashion and lifestyle is a matter of ethics, which makes it immensely important to become aware of the consequences of one's actions and choices in every single detail throughout the process (Plannthin 2016).

The fundamental and most elementary choice is to make a conscious choice, whether it resides in a design process or a manufacturer, and to examine this choice for further enlightenment. Therefore, the first decision to make is whether one is convinced that the use of animals as a resource is ethically acceptable or not.

Drude-Katrine Plannthin (2016) makes an intelligent call not only on consumers but also on designers and producers to enhance animal ethics and welfare in future fashion productions. "*Even though guidelines on right or wrong would make everything much easier when making decisions, the public often has a personal opinion in relation to culture and traditions. These personal feelings, emotions, experiences, and educations do have an influence on choices made when dealing with the use of animals as products. This is why it is necessary for not only consumers, but also designers and producers, to set the agenda and improve ethics and animal welfare in the future production of fashion and lifestyle products.*"..."*Basically, manufacturers, designers, and consumers must begin to relate to the fact that a resource in industrial production is alive in the sense that*

---

[16]An example of this, in another field, is the Biotech Advisory Board set up by DuPont in order to consciously search for divergent views in the periphery that could help define a more robust strategy for the development of biotechnology. The company has striven to include various stakeholders from India, Africa and Latin America in its discussions. It has also invited environmental advocates, such as the former president of Greenpeace International to offer divergent views on the matter. The exposure of high-ranked managers and business leaders to radically opposite views has prompted changes and significant improvements in the company's approach and strategy for selling biotechnology. New ideas for future business models have emerged in line with the company's efforts to leave behind products based on petrochemical material and favor biologic-based businesses (Hart and Sharma 2004).

animals are living beings and must be treated differently as opposed to when growing and using plants or other man-made fibres. The first step for any producer, designer, or consumer must be to recognize and acknowledge this fact to act and develop an awareness and sense of responsibility in this area. It is necessary to be open minded and prepared to examine these issues, which can help in accessing and relating to future design processes, production methods, and daily consumption."

Animals are sentient beings, and it is the responsibility of humans to ensure that they have a "life worth living." The maltreatment of animals can cause severe reputational damages in relation to retailers, consumers, and other stakeholders. As set forth in the UNGC Code of Conduct Principle 11, both producers and suppliers should consider the fact that wild animals and their habitats are part of their natural ecosystems and should, therefore, be valued and protected and that their welfare must be taken into account.

Exhibit 1—NICE Code of Conduct Principle 11

Principle 11. We recognize the conscious ethical decision not to use real animal fur.

In businesses where animals are used for materials in production and/or labor, such animals must be treated with dignity and respect.

No animal must deliberate be harmed or exposed to pain. Taking the lives of animals must at all times be conducted by trained personnel using the quickest and least painful and non-traumatic methods available. These must be approved by trained veterinarians and conducted by competent personnel.

It is important to recognize and respect that animal have a mind and body, which can be harmed due to wrong, ignorant and brutal treatment.

Using animal products in fashion is a legitimate practice as long as it is recognized that animals are sentient beings. It is the responsibility of humans to ensure that the animals have a "life worth living."

Production, breeding and keeping of animals shall be conducted professionally securing the animals the right and sufficient food and water and this must consider and respect the individual animals physiological, health and behavioral (space, rest etc.) needs including in circumstances involving transportation of the animals. Wild living animals shall never be captured and used.

Any use of endangered animal species as defined by the Convention on the International Trade in Endangered Species is strictly prohibited and the relevant authorities will immediately be contacted in such regard.

An example of animal care policy is shown below:

<COMPANY> regards the protection and respect of animals as very important. The supplier and subcontractors warrant that they adopt the same view and stance.

The supplier and subcontractors must submit documentation in writing with photographs and/or video footage of the daily handling of animals involved as labor or production in the full production line. Situations in which the animals are at risk of being in pain (due, for instance, to attacks from insects and diseases, or when their life is being taken) must be submitted to the <COMPANY> in writing with a thorough description of the used method,

available methods, business standards and why the chosen method is used and which efforts
are considered to reduce pain to the animals.

Exhibit 2—PETA's Wool Video

PETA has shown us video footage from within the Ovis 21 farm network that supplies
merino wool for Patagonia's base layers and insulation. It is as disturbing as anything
PETA puts out. Four minutes long, the video contains graphic footage depicting
inhumane treatment of lambs and sheep, of castration, of tail docking (the removal of a
sheep's tail), and of slaughter of lambs for their meat. See more details below.

It is especially humbling to acknowledge responsibility for the practices shown
because our original involvement in this project was intended for, in addition to
restoring grassland, improving animal welfare. In 2005, we became aware of
(through PETA's campaign against Australian wool growers) the painful practice of
mulesing sheep to reduce the damage from flystrike. We worked to stop sourcing
wool on the open (and untraceable) market as quickly as we could, and even
delayed a major product launch of merino base layers until we could find reliable
sources for non-mulesed wool in New Zealand and Australia.

PETA has targeted Patagonia because it held us responsible for practices done in
our name: wool from the farms shown in the video is spun, knit and sewn into
clothing that bears our label. We accept responsibility for everything done by our
suppliers at any level, but especially in this case. Beginning in 2011, we embarked
on a close partnership with Ovis 21 to develop a radical new way to grow wool—
one that regenerates rather than depletes grassland, keeps alive a way of life in the
Patagonia region, and produces wool of unprecedented quality for our next-to-skin
clothing. This has been a significant and engaging project for us.

When we began our initial discussions with Ovis 21, we were happy to learn that
blowfly does not inhabit Patagonia, so mulesing is not an issue there. We were also
pleased to learn the Ovis 21 farmers took steps to ensure that animals have sufficient
fleece to maintain warmth through the winter. In addition, Argentina does not
permit the export of live sheep, a dangerous practice. And we noted that to achieve
certification from Ovis 21, participating ranchers must adhere to strict protocols for
grazing and land management, flock improvement, and shearing; all of which
favorably influences animal welfare. We have worked closely with Ovis 21 on its
progress toward holistic grazing; however, beyond verifying that no mulesing
occurs, we have not audited its animal-welfare practices and were unaware of the
issues raised in the video.

PETA does not believe in the use of animals for any human purpose; this is a belief
we respect but do not share. Nevertheless, PETA plays an important role in raising
awareness of harmful practices involving animals, and we listen when legitimate
concerns are uncovered, even if we become a target of their activism.

For our part, we do offer alternatives to down and wool for our vegan friends and
customers. In addition, we have allocated considerable resources toward the
development and implementation of the world's most stringent standard to ensure
that goose and duck down come from animals that have been neither live-plucked

nor force fed to produce *foie gras*. Our requirements were incorporated by the independent certification body NSF International into its Global Traceable Down Standard.

We have also been working on wool. In early 2014, we began working collaboratively with numerous other brands and the Textile Exchange to develop the forthcoming Responsible Wool Standard for treating sheep and lambs that meets 21st century moral standards for the ethical treatment of animals. It is our hope that this global standard, when completed, will protect animal welfare, influence best practices, ensure traceability, and ultimately give consumers clear and trustworthy information that will allow them to make responsible choices. PETA was invited by the Textile Exchange to join this process, but declined. The process did include the participation of other animal-welfare organizations.

It should be noted that two practices highlighted in the video are standard across the wool industry, for good reason. Castrating select members of the flock helps keep it manageable and eliminates overcrowding, while tail docking reduces instances of infection in sheep and facilitates hygiene. It is critical that these procedures be done humanely, in a way governed by enforceable, uniform standards.

We are not immune to shocking images. There is no excuse for violent shearing methods and inhumane slaughter. We are investigating the practices shown. We will work with Ovis 21 to make needed corrections and improvements, and report back to our customers and the public on the steps we will take.

We apologize for the harm done in our name; we will keep you posted.

Timeline of Patagonia's efforts to build a more responsible wool supply chain

February 2005
We learn about the painful mulesing process as a result of a PETA campaign against Australian wool growers, which also decries the "live export" of animals (shipping and selling of live sheep from Australia to the Middle East for fresh consumption).

2008
We begin to move our wool fiber source from purchase on the conventional open market, where wool is untraceable and mulesed, to non-mulesed regions in New Zealand and certain specific Australian supply chains where the practice does not occur. This requires delaying the introduction of our wool base layer program until we had a traceable supply chain in New Zealand and Australia where we could be sure mulesing was not used.

Fall 2008
We launch our Merino Performance Baselayer line, sourced from non-mulesed sheep in New Zealand.

November/December 2011
Our materials and environmental teams visit Ovis 21 network farms in the Patagonia region of Argentina to determine the viability of nominating yarn from

their program in order to (a) support an important grasslands conservation effort, and (b) maintain our policy of avoiding the mulesing of sheep in our supply chain.

Because blowfly infestations do not occur in Argentina, we confirmed mulesing is not occurring but did not audit the farms specifically for other animal welfare concerns. We are aware that tail docking, which reduces instances of infection in sheep and facilitates hygiene, is occurring. We do not explore castration practices.

Fall 2011
We move our Merino Performance Baselayer program to Australian traceable, non-mulesed wool.

2012
We continue with planning, quality testing, volume, and supply chain trials surrounding the Ovis 21 wool—adopting their fiber into more of our products over time.

Fall 2012
We introduce Ovis 21 wool in all of our socks and some baselayers.

February 2014
We begin work as part of a public task force on the industry approach to a Responsible Wool Standard (RWS), led by the Textile Exchange. The standard will ensure a responsible, consistent approach to treating sheep and lambs that meets 21st century moral standards of the ethical treatment of animals. It is our hope that this global standard, when completed, will protect animal welfare, influence best practices, ensure traceability, and ultimately give consumers clear and trustworthy information that will allow them to make responsible choices.

For Patagonia and the Ovis 21 network, RWS will emphasize animal welfare as a clear priority alongside grasslands restoration.

2014
We convert the entire Merino Performance Baselayer line to Ovis 21 wool.

We begin moving to correct the small part of our wool that is still bought on the open market (some wool hats, and the wool lining of our wetsuits) by changing suppliers.

Winter/Spring 2015
Through our Social and Environmental Responsibility team, Patagonia continues to help and lead the industry effort to develop the Responsible Wool Standard—participating in the Textile Exchange Working Group Steering Committee.

Summer 2015
We hold several internal meetings to decide how to begin the work of implementing the Responsible Wool Standard in anticipation of its completion in 2016.

August 2015
PETA shows us a video containing graphic footage depicting inhumane treatment of lambs and sheep, of castration, of tail docking, and of the slaughter of lambs for their meat.

While we previously understood the need to adopt a strict standard to ensure animal welfare and worked toward that goal, we were not aware of any animal welfare issues with Ovis 21 farms until now. We begin an urgent investigation into the practices shown in PETA's video and commit to working with Ovis 21 to make needed improvements, reporting back to our customers and the public on steps we are taking.

Note: We will monitor the video link and update as necessary.

Overview of Ovis 21 protocols involving animal welfare

To be certified as Ovis 21 Sustainable Wool, ranchers must adhere to strict protocols for grazing and land management, flock improvement, and shearing—three pillars that all include major provisions to ensure animal welfare.

In short, these standards stipulate:

- Sheep are bred in natural grasslands
- There is no mulesing
- No antibiotics or hormones are used
- Medical treatments are limited to vaccines and pyrethroids to control external parasites
- There is no "live export" of animals
- Castration and tail docking, industry standard practices that promote responsible flock management and hygiene, are done at an early age with techniques designed to minimize pain.

In detail, the three pillars contain the following provisions:

1. GRASS Standard

Land management is a key issue for animal welfare. With our adaptive management, and mainly by using holistic planned grazing, we achieve the following outcomes:

- Produce under extensive open paddocks, in conditions that mimic natural grazing and recreate the herbivore-predator relationship (no confinement or artificial feeding)
- Increase the forage available, and ensure that every animal has enough feed for the whole year
- Increase grassland biodiversity, which improves the quality of the diet
- Increase managers' skills and attention to be responsible for the animals
- Risk reduction and better management in case of drought or heavy snow that may cause animal losses

- Better water supply and distribution
- Use of guard dogs to help avoid predation by foxes and pumas in a friendly manner with the predator.

2. Flock Improvement Standard

- Breed open face, plain-bodied animals, which are better adapted for extensive grazing systems. Open faces ensure clear vision throughout the year, and correlate with fertility and fitness
- Breed for high fat deposition that correlates with the ability to survive and breed under climatic stressful conditions
- Breed solely using natural methods (no artificial gene manipulation)

3. Wool Classing and Shearing Standards

- Comply with Argentina's National Standard in all procedures involving shearing, classing and packing of wool (above the standard for advanced flocks)
- Shear using snow combs or blades to reduce cold stress and leave more wool on the skin in order to increase insulation

Exhibit 3—Patagonia CEO Statement
Patagonia to Cease Purchasing Wool from Ovis 21
   Dear Friends,
   We've spent the past several days looking deep into our wool supply chain, shocked by the disturbing footage of animal cruelty that came to light last week. Patagonia's partnership with Ovis 21 has been a source of pride because of the program's genuine commitment to regenerating the grassland ecosystem, but this work must come equally with respectful and humane treatment of the animals that contribute to this endeavor.

   The most shocking portion of PETA's video shows the killing of animals for human consumption. Like those in the Ovis 21 network, most commercial-scale ranches that produce wool from sheep also produce meat. What's most important is that we apply strong and consistent measures to ensure animals on ranches that supply wool for products bearing the Patagonia name are treated humanely, whether during shearing or slaughter. We took some important steps to protect animals in partnering with Ovis 21, but we failed to implement a comprehensive process to assure animal welfare, and we are dismayed to witness such horrifying mistreatment.

   In light of this, we've made a frank and open-eyed assessment of the Ovis program. Our conclusion: it is impossible to ensure immediate changes to objectionable practices on Ovis 21 ranches, and we have therefore made the decision that we will no longer buy wool from them. This is a difficult decision, but it's the right thing to do.

Re-building our wool program—with a partner that can ensure a strong and consistent approach to animal welfare, while also fostering healthy grasslands—will be a significant challenge. However, we reject the notion that cruelty is essential to wool production, despite what PETA claims. Patagonia will continue to make products from wool because of its unique performance attributes. We will continue to sell products made from the wool we've already purchased. And we will continue to offer excellent synthetic alternatives for those who prefer them, while constantly pushing to innovate and invest in new materials and better supply chains. But Patagonia will not buy wool again until we can assure our customers of a verifiable process that ensures the humane treatment of animals.

We will also continue our efforts, initiated in 2014, to lead in the industry's development of strong, new verifiable standards for wool production we can all be proud of. We will take this as an opportunity to push even harder for the strongest possible animal welfare standards to be integrated into the forthcoming Responsible Wool Standard.

We apologize for the harm done in our name. We will continue to update you on our progress to do better.

Rose Marcario
CEO Patagonia

Exhibit 4—Stella McCartney in Instagram
Post 1—August 13, 2015[17]

> I am very saddened to report that we have had to cease sourcing some of our sustainable wool from Ovis 21 in Patagonia. It was born as an amazing initiative to help protect a million acres of endangered grasslands in Patagonia whilst looking after the welfare of animals. Unfortunately, after conducting our own investigation in Argentina, following a very distressful viewing of footage provided by the great guys at @officialpeta, we found out that 1 of the 26 ranches we used source sustainable wool there, mistreated its sheep. It is one too many.

Post 2—August 13, 2015[18]

> As a designer who built a brand on not using leather, fur or animal skins in its designs, I can't tolerate it! I am devastated by the news but more determined than ever to fight for animal rights in fashion together and monitor even more closely all suppliers involved in this industry to end all innocent lives. We are also looking into vegan 'wool' as well, in the same manner we were able to develop and incorporate high-end alternatives to leather and fur over the years.

---

[17]Source https://www.instagram.com/p/6V4FQyLmBL/.
[18]Source https://www.instagram.com/p/6V8KF3rmJT/.

Exhibit 5—Ovis 21's Statement

09:43
03/10/2016

Exhibit 6—Consumers' Reactions Posted on Patagonia Web site
Moana said[19]:

> Thank you for the strong stance. After viewing the video I have decided to discontinue my purchase of wool. I wonder if Patagonia would consider creating a sheep sanctuary that protects the grasslands.

Caitlin said[16]:

> Thank you so much. I happily pay more for Patagonia down and other products, because I want the highest quality with humane standards. I will do the same for wool. I look forward to seeing how Patagonia can change this industry.

Kathyn Gibson said[16]:

> I have made the decision to get rid of all my Patagonia product. I find it incredible and unbelievable that you had absolutely no visibility on your suppliers which tells me you have no quality control programs in effect. Really? I choose to not buy any more of your merchandise and will do everything I can to further educate others in your tacit approval and acceptance of your calculated indifference and feigned ignorance relating to your supplier. Shame on you, how can you sleep at night. Coward liars.

---

[19]Source http://www.thecleanestline.com/2015/08/patagonia-to-cease-purchasing-wool-from-ovis-21.html. Accessed: October 1, 2015.

GildaLee27 said[20]:

Based on this video, I immediately wrote to Patagonia (including the company's CEO, Rose Marcario) expressing my disapproval and requesting Patagonia's public acknowledgement and apology for using Ovis 21, the cruel and inhumane supplier.

Lucy_P, said[21]:

Stella has integrity. I wish Patagonia would follow suit, instead of offering up lame excuses. Whenever animals are treated as commodities—instead of the thinking, feeling beings they are—abuse inevitably follows. Vague assurances mean nothing and happy sounding labels are a farce. The only humane materials are vegan ones. Thanks to PETA for exposing what really goes on in the wool industry!

Rachel-Erin Hendrix, said[22]:

The real rub is there are total ethical and humane ways to raise sheep for wool production! As a wool spinner and knitter I have met some wonderful and kind sheep herders! There is no excuse for this, big corporations don't care about the process, just the profits, and that is utterly disgusting.

References

Allwood, J. M., Laursen, S. E., Malvido de Rodríguez, C., & Bocken, N. M. P. (2006). *Well dressed? The present and future sustainability of clothing and textiles in the United Kingdom.* Cambridge: University of Cambridge, Institute for Manufacturing.

Animal Equality. (s.d.) Clothing: Supporting animal slaughter. http://www.animalequality.net/clothing. Accessed: 3 March 2016.

B Corp. https://www.bcorporation.net/. Accessed: 1 Mar 2016.

Born Free USA. (2014). Get the facts: Ten fast facts About fur. Retrieved from http://www.bornfreeusa.org/facts.php?blog=9&more=1&paged=3. Accessed: 1 Mar 2016.

Byars, T. (2015). PETA's wool video. Retrieved from http://www.patagoniaworks.com/press/2015/8/12/petas-wool-video Accessed: 20 Aug 2015.

Cavagnaro, E., & Curiel, G. (2012). *The three levels of sustainability.* Sheffield: Greenleaf Publishing.

Dickson, M. A., Loker, S., & Eckman, M. (2009). *Social responsibility in the global apparel industry.* New York: Fairchild Books.

Dobson, A. (2003). *Citizenship and the environment.* Oxford: Oxford University Press.

Draper, S., Murray, V., & Weissbrod, I. (2007). *Fashioning sustainability—A review of sustainability impacts of the clothing industry.* London: Forum for the Future.

Ehrenfeld, J. R., & Hoffman, A. J. (2013). *Flourishing—A Frank conversation about sustainability.* Sheffield: Greenleaf Publishing.

Fletcher, K. (2008). *Sustainable fashion and textiles—Design journey.* London: Earthscan.

---

[20]Source https://www.youtube.com/watch?v=N7nKvgaMEaU Accessed: October 1, 2015.

[21]Source http://fashionista.com/2015/08/stella-mccartney-peta-wool. Accessed: October 1, 2015.

[22]Source http://www.huffingtonpost.co.uk/2015/08/17/peta-investigation-stella-mccartney-patagonia-wool-supplier_n_8000144.html. Accessed: October 1, 2015.

Fuertes, F., & Goyburu, M. L. (2004). *El Perfil sobre las Comunicaciones del Progreso en Argentina – Qué Comunican las Empresas del Pacto Global?*. Buenos Aires: Oficina del Pacto Global en Argentina.

Gardetti, M. A., & Torres, A. L. (2011). Gestión Empresarial Sustentable en la Industria Textil y de la Moda. Paper presented at the VII Congreso Nacional de Tecnología Textil - La sustentabilidad como desafío estratégico. INTI Buenos Aires 3–5 Aug 2011.

Gordon, J. F., & Hill, C. (2015). *Sustainable fashion—past*. Bloomsbury, London: Present and Future.

Gwilt, A., & Rissanen, T. (2011). *Shaping sustainable fashion—Changing the way we make and use clothes*. London: Earthscan.

Hart, S. L., & Sharma, S. Y. (2004). Engaging fringe stakeholders for competitive imagination. *The Academy of Management Executive, 18*(1), 7–18.

Hughes B (1988) Welfare of intensively housed animals. The Veterinary Record, 1988 Oct 1;123 (14):378.

International Labour Organization. (2006). Global Employment Trends Brief. http://www.ilomirror.cornell.edu/public/english/employment/strat/download/getb06en.pdf. Accessed: 17 Oct 2006.

International Standard ISO 26000 (2010) Guidance on social responsibility - First edition [PDF document].

Kaway, M. (2009). *Corporate social responsibility through codes of conduct in the textile and clothing sector*. Geneva: Graduate Institute of International and Development Studies.

Kell, G. (2003). The global compact—Origins, operations, progress, challenges. *The Journal of Corporate Citizenship Issue, 11*, 35–49.

Kelly, K. (1999). *New Rules for the new economy*. USA: Penguin.

McIntosh, M., Waddock, S., & Kell, G. (2004a). *Global compact, special issue (11) of the journal of corporative citizenship*. Sheffield: Greenleaf Publishing.

McIntosh, M., Waddock, S., & Kell, G. (2004b). *Learning to talk—Corporative citizenship and the development of the UN global compact*. Sheffield: Greenleaf Publishing.

Molderez, I., & De Landtsheer, P. (2015). Sustainable fashion and animal welfare: Non-violence as business strategy. *Conflict Management, Peace Economics and Development, 24*, 351–370.

Nordic Fashion Association. (2012). http://www.nordicfashionassociation.com Accessed May 24, 2012.

Nordic Fashion Association and Nordic Initiative Clean and Ethical. (2012). NICE Code of Conduct and Manual for the Fashion and Textile Industry. Nordic Fashion Association, Copenhagen.

Paulins, V. A., & Hillery, J. L. (2009). *Ethics in the fashion industry*. New York: Fairchild Books.

People for the Ethical Treatment of Animals—PETA. (2015). Stella McCartney Cuts Ties With One of Her Wool Suppliers as PETA Video Exposes Routine Mutilations. Retrieved from: http://www.peta.org/blog/stella-mccartney-cuts-ties-with-wool-supplier-as-peta-video-exposes-routine-mutilations/. http://www.peta.org/blog/stella-mccartney-cuts-ties-with-wool-supplier-as-peta-video-exposes-routine-mutilations/ Accessed: 20 Aug 2015

Plannthin, D.-K. (2016). In S. S. Muthu & M. A. Gardetti (Eds.), *Green Fashion—* (Vol. 1, pp. 49–122). Singapore: Springer.

Presas, T. (2001). Interdependence and Partnership: Building blocks to sustainable development. international journal of corporate sustainability. *Corporate Environmental Strategy, 8*(3), 203–208.

Qz.com. (2015). http://qz.com/479984/patagonia-and-another-ethical-clothing-brand-are-being-accused-of-a-new-kind-of-animal-cruelty/. Accessed: 9 Sept 2015.

Ross, R. J. S. (2009). *Slaves to fashion: Poverty and abuse in the New Sweatshop*. Ann Arbor: The University of Michigan Press.

Slater, K. (2000). *Environmental impact of textiles—Production, Processes and Protection*. Cambridge: Woodhead Publishing Limited—The Textile Institute.

Snowdon, K. (2012). Peta Investigation Prompts Stella McCartney To Sever Ties With Patagonia Wool Supplier, Ovis 21. Retrieved from: http://www.huffingtonpost.co.uk/2015/08/17/peta-investigation-stella-mccartney-patagonia-wool-supplier_n_8000144.html. Accessed: 7 Sept 2015.

UK Department for Environment, Food and Rural Affairs -DEFRA. (2008). *Sustainable clothing roadmap briefing note December 2007: Sustainability impacts of clothing and current interventions*. London: DEFRA.

The Virtuous Circle: Hard Sustainable Science Versus Soft Unsustainable Science Within Marketing Functions of Fashion and Luxury Sectors and How to Prevent 'Soylent Green' from Happening

Chloé Felicity Amos, Ivan Coste-Manière, Gérard Boyer
and Yan Grasselli

Abstract Not merely a business function, the view that marketing is being informed from science is increasingly gaining gravitas. However, the notion of 'scientific marketing' seems somewhat paradoxical and in considering marketing as a science, one enters into a deeply subjective and intricate matter. In considering this notion from a simplistic standpoint, it is clear that there are two distinct categories in which marketing falls into. Strongly influenced from marketing research, and equally from the intuitively creative proposals that arise from such research, one can deduce that scientific marketing incorporates both hard and soft sciences. However, this is where such a complex notion arises. Considering of how marketing is informed by phenomena such as culture, management and personal preference and relating this to the multitude of disciplines held by science, such as biology, psychology, neuroscience and sociology; one begins to understand what a

C.F. Amos (✉)
Luxury and Fashion Management, SKEMA Business School, Lille, France
e-mail: chloefelicity.amos@skema.edu

I. Coste-Manière
Luxury and Fashion Management, SKEMA Business School, Sophia, France
e-mail: ivan.costemaniere@skema.edu

I. Coste-Manière
Luxury and Fashion Management, SKEMA Business School, Suzhou, China

I. Coste-Manière
Luxury and Fashion Management, SKEMA Business School, Raleigh, NC, USA

G. Boyer
Cosmetics Industry and Fine Chemistry, Université des Sciences Marseille, Marseille, France
e-mail: gerard.boyer@univ-amu.fr

Y. Grasselli
Environmental and Marine Sciences, SKEMA Business School, Lille, France
e-mail: yan.grasselli@skema.edu

© Springer Science+Business Media Singapore 2017   75
S.S. Muthu (ed.), *Textiles and Clothing Sustainability*, Textile Science
and Clothing Technology, DOI 10.1007/978-981-10-2182-4_3

vast and complex concept scientific marketing is. This piece seeks to analyse and clarify how marketing activities are informed by both hard and soft sciences in the light of the fashion and luxury sectors. This chapter will examine and evaluate how such complexities in scientific marketing aid and hinder this particular business function and then seeks to how the future of marketing in the fashion and luxury sectors will be informed by this scientific approach. The luxury and fashion industry caters to both sustainable and non-sustainable consumptions. In this chapter, we explain some definitions such as natural, organic, bio and synthetic. We examine what has led to the entropic syndrome from energy consumption to pollution, the real 'umweltschutz' has begun industry cases and interviews show the technicalities from production and consumption, enhancing some problematic elements with facts and data, dealing with problematic elements in clothing supply chain such as raw materials, water, energy, climate change, waste, chemicals emissions, disposal or biodegradation, on the one hand. On the other hand, a specific section of this chapter is analysing the industrial face of textiles and clothing, with references to the food, perfumes and cosmetics industries: industrial ecology in textiles, problematic areas and solutions. Definitely not a simple optimistic Stairway to heaven.

Keywords Marketing · Water · Energy · Climate change · Biodegradation · Textiles · Industrial ecology · Clothing supply chain · Raw materials · Chemical emissions

1 Introduction

No longer is marketing considered merely as a marketing function, the view that marketing is being informed by science is increasingly gaining gravitas. However, alongside this view there has been a long running debate regarding the scientific status of marketing. From a scholarly perspective, there has been a lack of substantial research bringing into question the credentials of the use of science within marketing. From a wider perspective, it may be that the reason for this lack of research is attributed to the fact that the notion of 'scientific marketing' seems somewhat paradoxical in nature. Considering marketing as a science, one enters into a deeply subjective and intricate matter.

Exploring scientific marketing from a simplistic standpoint, it is clear that it falls into two distinct categories. Strongly influenced from marketing research, and equally from the intuitively creative proposals that arise from such research, one can deduce that scientific marketing incorporates both hard and soft sciences. However, this is where such a complex notion arises. Considering of how marketing is informed by phenomena such as culture, management and personal preference and relating this to the multitude of disciplines held by science, such as biology, psychology, neuroscience and sociology, one begins to understand what a vast and complex concept scientific marketing is.

Coupling the notions contained within scientific marketing with sustainability, we are faced with a scenario which is as equally vast as it is relevant especially in the context of the luxury and fashion industries. It is clear that the fashion and luxury industry caters to both sustainable and non-sustainable consumption, especially in the light of its marketing processes.

This piece seeks to analyse and clarify how marketing activities are informed by both hard and soft sciences, with sustainability at its heart. This chapter will examine and evaluate how such complexities in scientific marketing aid and hinder this particular business function and then seeks to how the future of marketing will be informed by this scientific approach in a means which is sustainable.

2 Defining Marketing Science

Firstly, it is most appropriate to commence with defining marketing science in its mostly simplistic term, to then go on to analyse its vast and complex nature. Marketing science is recognised as a specific approach to marketing which concerns the fulfilment of customer needs and the development of techniques in which they are fulfilled predominantly through scientific methods. This is as opposed to more common marketing techniques based on humanities and arts. Scientific marketing is, moreover, concerned with the pursuit of truth within marketing on a vast scale.

Interest in this particular field has seen exponential growth within the last 4 decades; thanks to the use of electronic data collection, such as electronic points of sales, which has led to what has been described as a marketing information revolution (Brinker 2013). Although marketing has been long established to have a relationship with data information, with the increasing data that is being collected and the growing means in which it can be collected, marketing is relying on science for systematic means of observing, testing and measuring patterns which may be of use to marketing professionals. This is able to determine vast patterns in consumer behaviour, which can help to produce new business insights and build new strategic capabilities.

The objective use of data to support decision-making is illustrated within Fig. 1. Marketing as a science is about comprehending how to interpret data in the most effective means possible. It is further more about the ability to critically analyse data in terms of its quality, context and sources (Brinker 2013). Recognising gaps in data, minimising biases and making suitable judgements with imperfect information is what makes up marketing science.

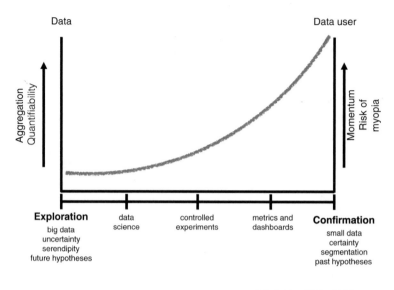

Fig. 1 Hard marketing science use

3 Multifaceted Nature of Marketing Science

It is important to recognise that what has been discussed is only a part of scientific marketing. What has just been explained can be considered as the analysis facets of marketing science, or the 'hard science', that informs marketing. It appears that the creative side of marketing (elements of branding, for example) is ignored as marketing science but this is very much a misconception. This creative side of marketing is what can also be referred to as the soft science of marketing. It is very clear that the analysis, which is used to inform the creative together make up the marketing science concept, incorporating both hard and soft sciences. Later within this chapter, we will go on determine the sustainability of both hard and soft sciences within marketing, respectively.

As we continue with the current theme, it is worth highlighting the challenging notion of marketing as a science. In examining this concept, especially in the light of sustainability, we are faced with some interesting, yet difficult elements to comprehend. As it has just been previously mentioned marketing is informed by both hard and soft sciences. These come under several and varied scientific categories as portrayed within Fig. 2.

These interact in a complex manner which makes it difficult to draw clear conclusions as what is hard science and what is soft science. Figure 3 illustrates 3 distinctive elements of marketing which are characterised by both hard and soft sciences (and a combination of the both) (Fig. 4).

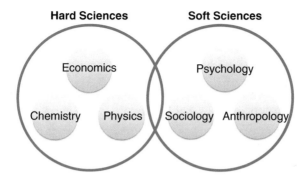

Fig. 2 Marketing—Hard and soft sciences

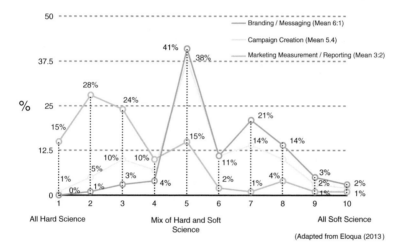

Fig. 3 Modern marketing—More hard science or soft science

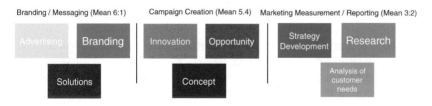

(Adapted from Brinker 2013)

Fig. 4 Ratio of hard to soft sciences regarding specific marketing activities

With these diagrams we begin to comprehend how the distinction between hard and soft sciences becomes somewhat 'fuzzy' and difficult to distinguish. We are able to successfully dispel the myth of marketing science being interpreted as a formulaically predictable phenomenon and is in actual fact a mix of both hard and soft sciences (Brinker 2013). However, in establishing this, we enter a more philosophical debate into the science of marketing. The hard dimension of marketing science devotes itself to the accurate measurement of empirical evidence, simply a mechanism for determining a means to an end (Brinker 2013).

The soft dimension of marketing science is a lot more difficult to define as it encompasses human psychology and behaviours, which interact in complex ways. Take culture, an anthropological science, as an example. It is clear that culture informs marketing decisions and strategic objectives of firms; one may proclaim more so than any other social phenomena. Culture, however, is such a complex notion to even define and to comprehend, let alone to allow it to inform marketing decisions. However, it is never the less a prominent component within marketing. One may even argue that it is the key. This is just one example of how soft sciences are difficult to form concrete conclusions and solutions for marketing means. This is not to say that they cannot interact with hard science means to provide marketing solutions. But, as Brinker (2013) states this interaction is usually on a relatively limited scope within specific scenarios. However, this cannot be an excuse to believe that such interactions between hard and soft sciences produce approaches which can match that of the actual knowledge of customer thinking and behaviour; at best they are simplified approximations within a narrow context (Brinker 2013). This notion will be expanded on, later within this chapter.

Thus far we have defined scientific marketing and it's highly complex nature. This is summed up perfectly by Hunt (1976), who defines it as 'intersubjective certification' which concerns the epistemologically uniqueness of results. However, in spite of such inherent difficulties, and this perfect sum up of such a tricky phenomena, there are definitely some prominent elements of scientific marketing that are worth exploring.

Firstly, there are some specific patterns within scientific marketing that can determine models and hypotheses for marketing purposes. But recognising the limitations within scientific marketing is nevertheless prevalent; it is ultimately about managing the complex system dynamics of marketing most effectively as possible (Brinker 2013). Secondly, it should be celebrated that marketing as a science utilises and embraces so many different phenomena from different scientific disciplines. Although translating these disciplines for use within marketing is challenging, it does provide endless inspiration for new marketing concepts. As previously mentioned, paying attention to relevance and reliability is still prevalent, but it does provide this 'business function' with a constant stream of new ideas, models and means to market successfully.

Overall, and pre-eminently, we can fully establish that, whether hard or soft, science informs marketing despite such complexities. There are specific guidelines that have been developed which have been adapted in Fig. 5, which illustrates how science, both soft and hard, can inform marketing in the most effective means.

These guidelines present how science in marketing can be effectively incorporated together to inform marketing decisions. Overall, science in marketing is a quest for knowledge, effective working knowledge that can be applied to an

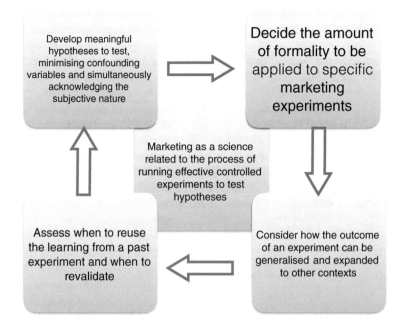

Fig. 5 Essence of marketing as a science taking into account hard and soft sciences

expanding business. However, as we move into the next section of this chapter, we will discover how sustainability influences marketing, and how soft sciences affect the sustainability of marketing.

4   Sustainability Within Marketing

There appears to be a prevalent problem at the heart of sustainable marketing that is as obvious as it is ubiquitous, a so-called value-action gap. Over the years, many sustainability campaigns have been appealing to our better intentions, values, ethics and so on (Townsend and Niemtzow 2015).

The reasons for why these campaigns were not as successful as hoped, relates to the emphasis of such campaigns. Focus should be placed not on values, but on value. The value offered by brands requires attention as most sustainably marketed products and services fail to answer the question 'what's in it for me?' (Townsend and Niemtzow 2015).

Understanding the benefit for customers is vital but often an overlooked and under explored component of sustainable marketing. By showing consumers what sustainability can do for them, rather than what they can do for sustainability, marketers can close the value-action gap a lot faster (Townsend and Niemtzow 2015).

This is especially prevalent with in the fashion and luxury industries. These sectors are increasingly indulging in this notion of sustainability. However, it is more centred around the idea of how sustainability can serve such consumers.

This is considerably more easily said than achieved. Sustainability seems to carry responsibilities not only related to what it's meaning implies, but also related to the idea of tapping into potential consumer value benefits. Being able to develop campaigns and marketing initiatives which fulfil both these requirements seems very difficult and demanding. If we are now to couple this with soft sciences within sustainable marketing, and the notion of being able to fully understand how consumers will react to sustainable marketing initiatives, we enter a very complex system.

5 Unsustainability of Soft Sciences Within Marketing

Earlier within this chapter, we touched on the complexity of soft science for the use of marketing purposes. We would like to expand on this further, encompassing the notion of sustainability. Culture, an anthropological science, was taken as an example which will be used again to illustrate how such a soft science in terms of marketing works together with sustainability. In order to do this, first we delve more into this anthropological science as a notion.

Culture should be viewed as dynamic open systems spread across geographical boundaries and evolve over time (Hong and Chiu 2001; Fang 2011). The ocean can be employed as a metaphor for the concept of culture. Fang (2005) states that, like culture, the ocean has no boundaries, contains various waters, both separate and shared, different and similar and dependent and independent. The metaphor emphasises how cultural variables can become more salient and rise to the surface while other may be temporarily suppressed, until awakened by a conditioning factor (Fang 2005, 2011). Critically, as a result of a multitude of difference variables, an array of hibernating values gives rise to the surface, thereby bringing about profound cultural changes into societies (Fang 2011). Although this metaphor gives and brings about a clear idea for this particular science, it nevertheless gives rise to a number of different issues which are all relevant from a marketing perspective. These issues are listed within Fig. 6.

Within marketing terms, this presents a very large issue in terms of how soft sciences are difficult to form concrete conclusions and solutions for marketing means. This is especially the case in terms of sustainability or unsustainability as we will discover.

One big issue which lies within sustainability and marketing is this notion of misguided solutions. Sustainability based marketing is usually only partly informed. In other terms, the sustainability element which is marketed upon usually

Issue	Problems
Definition	Lack of common definition on culture
Sampling	Opportunist sampling of cultures and non-representiveness of samples
Instrumentation and Measurement	Non equivalence of variables, nuances in definitions and non equivalent scaling
Data Collection	Researcher biases
Data Analysis and Interpretation	Ecological fallacy, qualitative versus quantitative data

Adapted from Lim and Firkola (2000)

Fig. 6 Issues and subsequent problems related to soft sciences (particularly anthropology: culture)

serves one particular interest, and only one sustainability problem is solved. However, from a wider perspective this can in actual fact potentially create other further problems. This is not more highly relevant than in terms of soft sciences. Soft sciences such as anthropology can misinform marketing due to such complexities that are associated with their nature. Sustainability is such a large issue within marketing but is seldom assessed due to the nature of soft sciences which informs it.

6 How Soft Sciences Such as Anthropology Misguide Sustainable Marketing

In order to propound this notion of how anthropology and sociology influence marketing decisions is a means which is not sustainable—marketing decisions which appear sustainable from a sociological perspective but are actually misguided, we present a new, specific framework. This framework aims to emphasise the arguably paradoxal idiosyncrasies within the concept of culture. It highlights how the ever-present challenge of bringing clarity to the conception of culture ultimately means unsustainability within marketing terms (Fig. 7).

Like culture, changes/movements in the kaleidoscope are reflected in an endless variety of patterns, which are all interlinked with a succession of changing phases. New perspectives and new changes will create new images, with both interesting and continually varying focal points. What is focused on within the kaleidoscope is subjective to the user/researcher. As this focal point is explored in more depth and the kaleidoscope is turned, subtle changes are created that change the overall perception of the kaleidoscope image. There is nevertheless a core image presented; however, facets within the complete picture are constantly developing and changing.

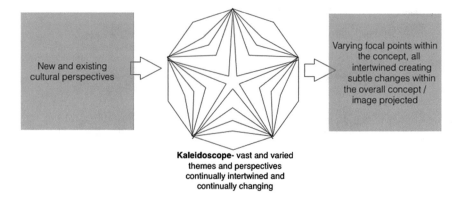

Fig. 7 Kaleidoscope model of soft sciences: focusing on culture (anthropology as a soft science)

This framework can be used as a means to inform marketers that the complex nature should be accepted and not necessarily challenged, as there are complex relationships within this soft science which are intertwined and ever evolving. It can, moreover, be used as a means to illustrate how such a concept ultimately results in misjudgements in terms of sustainable marketing (Fig. 8).

In presenting this particular framework, the intention is to provide an overall scope of how such a soft science can interrupt marketing procedures with so many nuanced areas which disturb this particular business function.

Perspectives are altered as a result of nuances within culture adjusting the reality of what is perceived and believed. Outcomes which may appear to have meaning and reasoning are not necessarily absolute. This ultimately results in misjudgements in terms of what sustainability actually constitutes. Moreover, and as it has been previously touched upon, this can actually act as a 'cover-up' for potentially bigger problems. This cannot only be detrimental to a product or service being marketed, but also to overall business, the brand, its reputation and profitability.

It is obviously very difficult to determine a single area in which misinterpretations may appear, and this is why such a framework is so relevant. Even if

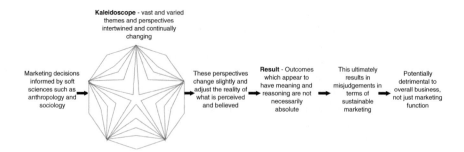

Fig. 8 Kaleidoscope model: soft science influence on marketing decisions

specificities were to be identified, they would have likely to have changed soon after being identified. But this is the nature of this soft science. It should be stressed that despite the inability to explicitly identify areas of specificity, there is nevertheless an interesting future for soft science in marketing, despite its unpredictable nature.

The purpose of using such a framework is also to emphasise the intricate nature of such a soft science within its entirety. Like the kaleidoscope, a myriad of differing shades, patterns and focal points are produced with every new turn/discovery. This makes it incredibly difficult from a marketing perspective being able to arrive at absolute marketing results. However, on the other hand, one may argue that this makes up a part of the essence of marketing. The unpredictability fuels the need for marketing. Furthermore, it provides endless inspiration for new ideas, models and marketing concepts.

7 How This Is Relevant to the Luxury Industry

Classically speaking, luxury has been very much associated with, and has revelled in, soft sciences, both as a social theory and also within business/marketing terms. Luxury appears to be a rewarding subject for soft sciences, in particular sociology for example and this is still somewhat prevalent today. According to Schrage (2012), classically, luxury has placed focus (to arguably some degree) on highlighting social exclusivity, which had rendered it effective beyond the circle of luxury consumers. There were distinctive behavioural patterns of luxury consumption that were used as signifiers for social status, for example (Schrage 2012). Marketers have profited from playing to these behavioural patterns to attract luxury clients and fix the notion of luxury with a distinct social outlook. One may proclaim that it is soft marketing at its peak, taking into account behavioural patterns and social cues to influence clients purchasing intensions. Classic luxury marketing encompasses human psychology and behaviours.

According to Schrage (2012) within modern luxury consumption, these behaviours and ideals are no longer as prevalent, and the landscape for luxury in terms of soft science marketing means is a little different, but slightly more complex. The types of clients are changing, the demands, also are changing. Therefore, the link between luxury and soft science marketing in a modern capacity has seemed to be shifted and it's social function is certainly not what it once was (Schrage 2012). Soft social sciences have turned their attention to the subtle distinctions within luxury buying behaviours. This is where the ideas of the kaleidoscope mode really come into force, illustrating, specifically within the luxury industry, how differing yet subtle distinctions within the 'social' elements of the sector making it difficult from a marketing perspective, being able to arrive at absolute marketing results. One may proclaim that this is the reason why branding, brand identity and marketing within the luxury industry are so unique, thought-provoking and boundary pushing, yet nevertheless challenging to maintain constantly.

Add in the notion of sustainability into the mix and picture becomes even more complex. Nevertheless, a sector in which sustainability is increasingly taking a central role, it is an important element to mention within the capacity of soft science marketing and the luxury industry. However, for the sustainability elements to hold true they must be portrayed in the most clear and concise manner and must serve for the greater good. For sustainability to be portrayed in this manner seems to contradict soft science marketing efforts. Nevertheless, it is a challenge worth overcoming within an industry heavily reliant of soft marketing and with luxury clients increasingly building an affiliation with the products consumed and the notion of sustainability.

8 Conclusion

Whether soft science causes the marketing to be unsustainable or influences its sustainability perception is a very general statement. From our analysis, it appears that soft science does seem to hinder marketing efforts. However, what is also apparent is that without such soft sciences, the need for marketing would be less great. Add in the notion of sustainability makes this vision unclear still, however, marketing sustainability, and the future of sustainable marketing is still and will continue to be reliant on soft sciences. As it has been explored in the latter of this chapter, the luxury industry is one at the forefront of this 'dilemma'. An interesting paradox, but extremely prevalent and arguably relevant for the foreseeable future. However, it is important to reiterate, that is, this paradox and this unpredictability that fuels the need for marketing in both the soft and hard forms. This is most relevant within the luxury sector and it will certainly be interesting to observe how these elves further in the future.

References

Brinker, S. (2013). *What do you mean by marketing as a science?—Chief marketing technologist.* Retrieved March 21, 2016 from http://chiefmartec.com/2013/03/what-do-you-mean-by-marketing-as-a-science/

Eloqua (2013). *Defining the modern marketing—From ideal to real. Eloqua.com.* Retrieved March 21, 2016 from http://demand.eloqua.com/modern-marketer?elqchannel=press-release

Fang, T. (2005). From "onion" to "ocean": Paradox and change in national cultures. *International Studies of Management and Organization, 35*(4), 71–90.

Fang, T. (2011). Yin Yang: A new perspective on culture. *Management And Organization Review, 8*(1), 25–50.

Hong, Y., & Chiu, C. (2001). Toward a paradigm shift: From cross-cultural differences in social cognition to social-cognitive mediation of cultural differences. *Social Cognition, 19*(3: Special issue), 181–196.

Hunt, S. (1976). The nature and scope of marketing. *Journal Of Marketing, 40*(3), 17.

Lim, L., & Firkola, P. (2000). Methodological issues in cross-cultural management research: Problems, solutions, and proposals. *Asia Pacific Journal of Management, 17*(1), 133–154.

Schrage, D. (2012). The domestication of luxury in social theory. *Social Change Review, 10*(2). http://dx.doi.org/10.2478/scr-2013-0017

Townsend, S., & Niemtzow, E. (2015). *The problem with sustainability marketing? Not enough me, me, me.* The Guardian. Retrieved March 21, 2016 from http://www.theguardian.com/ sustainable-business/behavioural-insights/2015/mar/09/problem-sustainability-marketing-not-enough-me

Social Reporting Using GRI Disclosures: A Case of Apparel Industry

Aswini Yadlapalli and Shams Rahman

Abstract In recent times, social responsibility has emerged as one of the critical issues for corporations globally. The importance given to social responsibility and sustainability in corporate board rooms is noticeable through sustainable reporting. One such reporting system is to report through the Global Reporting Initiative (GRI) framework. The purpose of this chapter is to provide an overview of the GRI framework and investigate the social reporting of apparel firms in the global context. Using a sample of 37 global apparel firms from the sustainability disclosure database who report GRI, a content analysis is performed and measured the level of firms' social disclosures. Results indicate that out of the four sub-categories of social disclosures, labour practices and decent work, and human rights are the most disclosed social sub-categories, whereas the society sub-category is least disclosed among apparel firms. Furthermore, the large firms are likely to publish sustainability reports compared to the small and medium enterprises. However, the size of the firm does not impact the level of social disclosure. Though the firms in the European region are more likely to publish sustainability reports compared to firms located in other regions, however, the level of their social disclosures are comparatively low.

Keywords Labour practices · Social reporting · GRI · Human rights · Apparel firms · Sustainability disclosure · Working conditions

1 Introduction

In 1990s, several high-profile apparel firms were involved in a number of sweatshop scandals, and in 2010s, several major incidents such as collapse of Rana Plaza and fire in Ali garment facilities and Tazreen Fashions have highlighted the social, safety, and employment issues at apparel manufacturing facilities (Moore et al. 2012).

A. Yadlapalli · S. Rahman (✉)
School of Business IT and Logistics, RMIT University, Melbourne, VIC 3000, Australia
e-mail: shams.rahman@rmit.edu.au

© Springer Science+Business Media Singapore 2017
S.S. Muthu (ed.), *Textiles and Clothing Sustainability*, Textile Science and Clothing Technology, DOI 10.1007/978-981-10-2182-4_4

This circumstance has forced the apparel brands to extend their responsibilities beyond their organizational boundaries (Perry and Towers 2013). As social responsibility became an important aspect for the apparel business, there is a growing dialogue in apparel community regarding sustainability issues (Kozlowski et al. 2012). In recent times, social responsibility has emerged as one of the top agenda items for corporations all over the world (Aras et al. 2010). The importance given to social responsibility sustainability in corporate board rooms is noticeable through sustainable reporting (Barkemeyer et al. 2015). One such reporting system is to report through the Global Reporting Initiative (GRI) framework. In the current business environment, many firms consider social and environmental reporting practices as an investment for sustainable development (Chen et al. 2015). The increased importance given to sustainability reporting can be not only seen in the number of reports published but also in the quality of the reporting. For example in 2011, over 80 % of the Fortune 500 companies and 95 % of the 250 global largest companies publish CSR or sustainability reports (Lii and Lee 2012; KPMG 2011); and the content of the reports has increased from just a page on sustainability to a detailed stand-alone sustainability reports (Qiu et al. 2014).

The objective of this chapter is to provide an overview of the GRI framework and investigate the social reporting of apparel firms in the global context. Further, this study examines the reporting content against contextual factors such as the firm size and location. The remainder of the chapter is organized as follows. Section 2 provides an overview of the apparel industry and its sustainable practices and sustainable reporting in general. The methodology adopted for sample selection and the procedure to calculate social disclosure scores are discussed in Sect. 3. Section 4 presents the results of the analysis. Finally, discussions on findings and conclusions are drawn in Sect. 5.

2   Apparel Industry & Sustainability

Apparel trade is considered as one of the world's leading merchandise trade with a share of 2.6 % in 2014 (WTO 2015). Over the past 50 years, global export of clothing and textiles has increased from under $6 billion in 1962 to $797 billion in 2014 (WTO 2012, 2015). Approximately, 70 % of the apparel is manufactured in developing nations. In the manufactured goods category, the textile and apparel industry has been rated as the single largest source of export income for the developing nations. This industry is driving the economic growth in many South and South-east Asian nations (Bhardwaj and Fairhurst 2010). The main reason for the trade shifts from developed nations to developing nations is the adoption of low-cost country sourcing strategy. Globalization in apparel industry along with the outsourcing of manufacturing activities to developing nations made the product affiliation to multiple countries and led to complex supply chains (Gereffi and Frederick 2010).

Table 1 Major incident in apparel manufacturing

Company, country	Year	Number of deaths	Cause
Triangle Shirtwaist Factory, USA	1911	100	Fire
Ali garment factory, Pakistan	2012	289	Fire
Tazreen Fashions, Bangladesh	2012	112	Fire
Rana Plaza, Bangladesh	2013	1132	Building collapse

In apparel supply chains, large retailers, marketers, or branded manufacturers lead the supply chain (Gereffi and Frederick 2010). In most cases, the lead retail firm outsources manufacturing process to a global network of suppliers. Huge competition among apparel manufacturers for foreign investments and contracts leave them with a little leverage. This context leads to an opportunistic behaviour among the supply chain members resulting in unacceptable practices along the apparel supply chains. Social issues in the apparel industry are classified into three major areas: wages, working hours, and working conditions. The industry is accused of paying low wages to workers, using underage and child labour; discriminating against gender, religion, and social class; abusing of human rights; introducing acts preventing employees to join unions; and failing to offer minimum labour standards (Awaysheh and Klassen 2010). These unethical practices resulted in major disasters in this industry. Table 1 provides the list of incidents in apparel manufacturing. The consequences of these devastating incidents questioned the unacceptable working conditions such as fire safety and structural integrity of the apparel factories.

The apparel industry is characterized by labour-intensive production and limited automation, competitive pressure to lower production costs, and transparency issues with several subcontractors (Park-Poaps and Rees 2010). High product variety, high volatility, low predictability, seasonality, and intense competition are the other factors that greatly influence the industry (Perry and Towers 2013). For instance, retail buying practices such as shorter lead times and variability in order quantities with demand for lower prices will enforce garment workers to work overtime without proper remuneration (Ruwanpura and Wrigley 2011).

To maintain a safe workplace and to manage social initiatives at manufacturing facilities, large multinational retailers along with NGOs have developed 'The Accord on Fire and Building Safety' and 'Alliance for Worker Safety'. Though the Accord and Alliance has been developed in Bangladesh context, soon it will be rolled over to the other manufacturers in developing nations. Likewise, several non-profit organizations have developed methodologies such as PAS 2050 and WRAP to measure the environmental and social impact of the apparel manufacturer and to accredit them. Most of these standards are considered as tools for the manufacturers to exhibit their responsibility towards employees and society.

The GRI framework is one of the most commonly used guidelines to report social and environmental initiatives of an organization (Kozlowski et al. 2012). GRI has also extended industry-specific guidelines known as Apparel and Footwear

Sector Supplement (AFSS) specifically designed for the apparel industry. Also, sustainable apparel coalition has developed indicators to measure and communicate the outcomes of sustainable initiatives. In this chapter, sustainability reports based on the GRI index is used to understand the social initiatives of global apparel firms. The following section describes the GRI framework in the light of the measures of social disclosures.

3 Sustainability Reporting

Patten (1992) referred social disclosure 'as a means of addressing the exposure companies' face concerning the social environment' (p. 297). Likewise, Pérez (2015) defined CSR reporting as 'the disclosure of company initiatives that demonstrate the inclusion of social and environmental concerns in business operations and interactions with stakeholders' (p. 11). The World Business Council for Sustainable Development explains that sustainability reports are 'public reports by companies to provide internal and external stakeholders with a picture of the corporate position and activities on economic, environmental and social dimensions' (WBCSD 2002).

The issue of sustainability is contextual based. It details on how a company is related to the environment in which it operates (Ellerup and Thomsen 2007). Hence, the content of the sustainability report depends on the factors such as the size of the firm, the complexity of the issue, the ambition level, the type of firm's engagement with society, and also the time period (Patten 2002). In general, the statements expressed in corporate disclosures signal the markets on how organizations are proactively managing social and environmental aspects (Chen et al. 2015). Organizations report sustainability initiatives either on their corporate websites or on their publishes as part of their annual reports and/or stand-alone sustainability reports (Kozlowski et al. 2012). Sustainability reporting contains a mix of qualitative and quantitative indicators. Organizations select the indicators that best suits to their context (Ellerup Nielsen and Thomsen 2007). There is no established framework for selecting indicators for communication; only guidelines are available (Patten and Zhao 2014).

Some of the widely recognized sustainability reporting standards are the GRIs G3 standards, AccountAbility's AA1000 Series, and the United Nations (UN) Global Compact's Communication on Progress (COP) (Tschopp and Huefner 2015). Among these standards, GRI provides guidance on reporting sustainability aspects that can be used across all sectors (Kozlowski et al. 2012). GRI frameworks are increasingly used to report sustainability initiatives on an annual basis in apparel firms to build a positive image and attract customers (Turker and Altuntas 2014).

GRI as an NGO is perceived as a credible organization to develop guidelines for non-financial information reporting (Knebel and Seele 2015). It emerged as a key governing body in sustainability reporting (Barkemeyer et al. 2015). Today, the GRI guidelines are regarded as 'the de facto global standard' (KPMG 2011, p. 20), are considered as a solution to the inconsistencies in reporting and can be used to improve the quality of reporting (Colicchia et al. 2013). A GRI report consists of three parts: first part relates to the profile of firms; part two is regarding management approach to sustainability, and part three is about the performance indicators. Among performance indicators, there are core indicators and additional indicators in economic, environmental, and social categories. Economic (7 cores and 2 additional), environmental (18 cores and 12 additional), and social (31 cores and 14 additional) indicators can be used in a sustainability report based on the organizations' intention of completeness. The social category of GRI indicators concerns about the impact of organization on social systems in which it operates. The social category is further divided into four sub-categories such as labour practices and decent work, human rights, society, and product responsibility. The social sub-categories are described as follows:

3.1 Labour Practices

Labour practices and decent work sub-category is concerned about the organizational employment practices and their relationship management with the employees. It also covers the aspects of workforce development and employee opportunities to experience the diverse and safe work place. The five aspects of labour practices are employment, labour/management relations, occupational health and safety, training and education, and diversity and equal opportunity (see Fig. 1). These aspects are explained as follows:

(a) Employment covers the total workforce and their turnover by region, gender, age, and contract. It also highlights the differences in the benefits of full-time and part-time/contract employees.
(b) Labour/management relations report on the percentage of employees covered by bargaining agreements and the terms of barging agreements related to notice periods.
(c) Occupational health and safety aspect deals with the percentage of health and safety topics covered in formal agreements; percentage of the employees represented in formal joint health and safety committees; and rate of work-related injuries, fatalities, and abscentism. It also covers the education and training of the employees on occupational health and safety.
(d) Training and education aspect reports on the skills management, carrier development, and lifelong learnings of the employees. It is measured based on the percentage of employees receive training and number of training hours per year.
(e) Diversity and equal opportunity aspect covers the composition of governance bodies and breakdown of employees based on gender, age, group, minority groups, and other indicators of diversity. It also covers the measures on ratio of basic men salary to women salary by employee category.

Category	Sub-Category	Aspect	Indicators

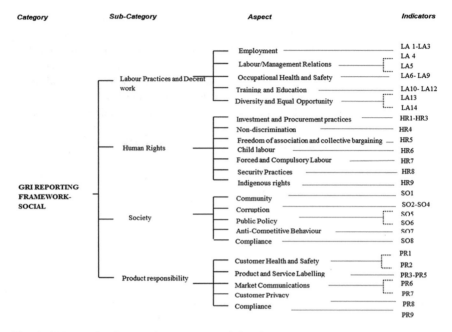

Fig. 1 GRI reporting framework to measure social performance

3.2 Human Rights

Human rights sub-category covers the extent to which an organization addresses the incidents related to human rights violations. It also reports the indicators related to the employee training on the awareness to exercise their rights. In addition, it emphasizes on the organization's consideration of human rights in investment and supplier/contractor selection. Investment and procurement practices, non-discrimination, freedom of association and collective bargaining, child labour, forced and compulsory labour, compliance and grievance practices, security practices, and indigenous rights are the seven aspects of human rights, which are further described below:

(a) Investment and procurement practice aspect reports on the percentage of the supplier contracts with human right clauses and suppliers that have undergone screening for human rights. It also reports on the total hours of employee training on policies and procedures concerning the aspects of human rights.

(b) Non-discrimination aspect reports on the total number of incidents of discrimination in an organization and the actions taken against it.

(c) Freedom of association and collective bargaining identifies the operations in which the right to exercise freedom of association and collective bargaining may be at significant risk. It also covers the actions taken against to support these rights.

(d) Child labour aspect reports on the operations at significant risk with the child labour incidents and the measures taken to eliminate the child labour.
(e) Forced and compulsory labour reports on the operations at significant risk with the incidents of forced and compulsory labour and the measures taken to eliminate the forced and compulsory labour.
(f) Security practices report on the percentage of security personnel trained in the organization's policies or procedures concerning aspects of human rights that are relevant to operations.
(g) Indigenous rights aspect reports on the total number of incidents involving violations of indigenous people rights and actions taken.

3.3 Society

The society sub-category concerns about the organizational impact on society and local community. The rights of the people under this sub-category are fundamental rights which are both collective and individual. The aspects of society sub-category are community rights, rights against corruption, public policy, anti-competitive behaviour, and compliance. These aspects are further explained below:

(a) Community aspect highlights on nature, scope, and effectiveness of programs and practices that assess and manage the impacts of operations on communities, including entering, operating, and exiting.
(b) Corruption aspects report on the total number of business units that are at risk due to corruption and the actions taken in response to the incidents. It also reports on the percentage of employees trained in organization's anti-corruption policies and procedures.
(c) Public policy covers the position of the organization and employees in public policy development and lobbying. It also presents the value of financial contributions to political parties, politicians, and related institutions by country.
(d) Anti-competitive behaviour reports on the total number of legal actions for anti-competitive behaviour, antitrust, and monopoly practices and their outcomes.
(e) Compliance aspect presents the monetary value fines for non-compliance with laws and regulations. It also reports on the non-monetary value of non-compliance with laws and regulations.

3.4 Product Responsibility

The last sub-category of the social category is product responsibility. Product responsibility sub-category deals with the issues of the direct impact of products and services on stakeholder and customers in particular. There are five aspects of product responsibility and they are customer health and safety, product and service labelling, marketing communications, customer privacy, and compliance.

(a) Customer health and safety covers the incidents regarding the impact of product and service life cycle on customer's health and safety. It also reports on the percentage of significant products and services categories subject to such procedures.

(b) Product and service labelling emphasizes on the percentage of the products; services require label information; and the non-compliance issues related to labelling requirements. It also reports on the practices related to the customer satisfaction measures.

(c) Marketing and communication indicator measures the programs that adhere to laws, standards, and voluntary codes related to marketing communications, including advertising, promotion, and sponsorship. It also reports on the total number of incidents of non-compliance with regulations and voluntary codes concerning marketing communications, including advertising, promotion, and sponsorship.

(d) Customer privacy aspects reports on the total number of substantiated complaints regarding breaches of customer privacy and losses of customer data.

(e) Compliance aspect covers the monetary value fines for non-compliance with laws and regulations concerning the provision and use of products and services.

All these sub-categories and aspects are developed based on the several recognized international standards, including United Nations Conventions, International Labour Organization (ILO) Conventions, and Organisation for Economic Co-operation and Development (OECD) guidelines. This chapter investigates the social reporting of apparel firms in the global context using the social category of the GRI framework. The economic and environment reporting is beyond the scope of this chapter. Figure 1 illustrates the social categories and their sub-categories used in this chapter.

4 Selection of Firms and Measures of Social Performance

As mentioned earlier, GRI guidelines provide a comprehensive reporting system for environmental and social performances. Therefore, we use the Sustainability Disclosure Database published by GRI for choosing the sample firms. The following criteria are applied to select the firms. The firm must

- belong to the apparel industry and
- publish standard GRI reports following sustainability reporting guidelines either version 3.1 or version 4.

Among 56 apparel companies listed in the Sustainability Disclosure Database, 37 firms satisfy the two selection criteria. Table 2 presents the demographic profile of these firms. It is clear that the sample includes companies across the world from countries with and without mandatory reporting. From Table 2, it is evident that the German apparel firms are leading in sustainable reporting (6 firms), followed by Switzerland (5 firms), and USA (4 firms). Apparel firms in other European countries such as Finland and Netherlands are also publishing sustainability reports significantly. Results demonstrate that the Europe is leading in sustainability reporting of apparel firms, followed by North America. The number of publications from European nations is consistent with the Wensen et al. (2011) global sustainable reporting that European firms are the leaders in sustainability reporting.

Table 2 also highlights that the sample firms are predominantly large (86.5 %), and few SMEs (13.5 %). In this study, a large firm employs more than 250 employees and had a turnover of greater than 50 million euros (86.5 %). The classification of firm's size is based on the EU definition as explained in Table 3.

4.1 Measures of Social Performance

GRI reports of the sample firms are content analysed to measure the quality of firms' social disclosures. An indexing technique is considered as an effective tool to gauge the level of disclosure (Bewley and Li 2000; Cho and Patten 2007). Each performance item in the GRI framework is measured using a grading scale of 1–5; where 1 indicates 'not reported', 3 indicates 'partially disclosed', and 5 indicates 'fully disclosed'. For example, a fully reported item may include quantitative measures, categories, and targets. The score for each category and sub-category of the firm is evaluated based on the Eq. (1) (see e.g. Bewley and Li 2000; Cho and Patten 2007; Zeng et al. 2010; Meng et al. 2014).

$$LD_i = \sum_{(j=1)}^{n} \text{Iij}/\text{nj} \tag{1}$$

where LD_i is the total score of labour disclosure for the firm i; Iij is the score of the jth term for the firm i; and n is the total number of j indicators in which $j = 1, 2, 3…$ 14 for the labour practices and decent work sub-category. In the similar manner, scores for human rights, product responsibility, and society sub-categories are calculated. An average of labour practices and decent work, human rights, product responsibility, and society scores gives social score of each firm.

Table 2 Demographic profile of sample firms

Company	Size	Region	Country
A&E	Large	North America	USA
Aksa Akrilik	Large	Asia	Turkey
ASICS	Large	Asia	Japan
CALIDA	Large	Europe	Switzerland
Crocs	Large	North America	USA
Crystal Group	Large	Asia	China, Hong Kong
CWS-boco Group	Large	Europe	Germany
DBL Group	Large	Asia	Bangladesh
Desso Holding B.V.	Large	Europe	Netherlands
Everest Textile	Large	Asia	Taiwan
Gildan	Large	North America	Canada
Hugo Boss AG	Large	Europe	Germany
HUMANA Kleidersammlung	SME	Europe	Germany
Impahla Clothing	SME	Africa	South Africa
Kering	Large	Europe	France
Lindström Oy	Large	Europe	Finland
Lojas Renner S.A.	Large	Latin America, Caribbean	Brazil
Mammut	Large	Europe	Switzerland
Mango Group	Large	Europe	Spain
Marc O'Polo	Large	Europe	Germany
Marimekko	Large	Europe	Finland
Marisol S.A.	Large	Latin America, Caribbean	Brazil
Memteks	Large	Asia	Turkey
Nike	Large	North America	USA
ODLO Sports Group	Large	Europe	Switzerland
Prada	Large	Europe	Italy
Puma	Large	Europe	Germany
PVH Corp.	Large	Northern America	USA
Remei	SME	Europe	Switzerland
Suominen Nonwovens	Large	Europe	Finland
Switcher	SME	Europe	Switzerland
TAL Apparel Limited	Large	Asia	China, Hong Kong
Teijin	Large	Asia	Japan
Toray Industries Inc	Large	Asia	Japan
VAUDE Sport GmbH & Co.KG	Large	Europe	Germany
Venus Colombia	SME	Latin America, Caribbean	Colombia
Zeeman	Large	Europe	Netherlands

Table 3 Classification of organizations based on the size (source: GRI 2015)

Enterprise category	Headcount	Turnover OR	Balance sheet total
SME	<250	≤€50 million OR	≤€43 million
Large enterprise	≥250	>€50 million OR	>€43 million

5 Analysis and Results

5.1 Social Indicators

As mentioned earlier, the social category is consisted of four sub-categories such as labour practices and decent work; human rights; society, and product responsibility. The level of disclosure about the sub-categories is shown in Fig. 2. Among the apparel firms, labour practices and decent work is the most often disclosed sub-category with a weight = 2.77 and society is least disclosed sub-category with a weight = 2.03 (see Fig. 2). The sub-categories are further analysed:

5.1.1 Labour Practices

The labour practices sub-category is consisted of five aspects and 14 indicators. Five aspects are employment; labour/management relations; occupational health and safety; training and education, and diversity and equal opportunity. The employment (weight = 2.91) is considered as the most critical aspect followed by diversity and

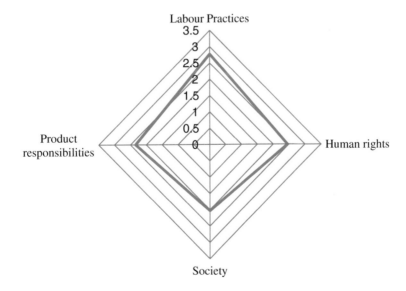

Fig. 2 Average scores of social sub-categories

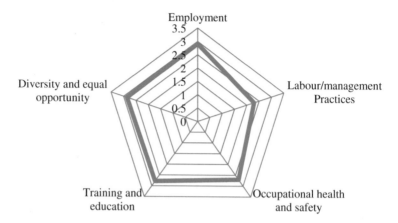

Fig. 3 Scores of labour practices and decent work sub-category

equal opportunity (weight = 2.86); training and education (weight = 2.78); and occupational health and safety (weight = 2.70). The labour/management relations (weight = 2.30) aspect is the least referred aspect under the labour practices and decent work sub-category (Fig. 3).

The top 3 indicators for labour practices are LA1: total workforce by employment type, employment contract, and region broken down by gender (weight = 3.76); LA13: composition of governance bodies and breakdown of employees per employee category according to gender, age group, minority group membership, and other indicators of diversity (weight = 3.32); and LA7: rates of injury, occupational diseases, lost days, and absenteeism, and total number of work-related fatalities, by region and by gender (weight = 3.22). LA6: Percentage of total workforce represented in formal joint management–worker health and safety committees (weight = 2.24) and LA9: health and safety topics covered in formal agreements with trade unions (weight = 2.30) are the two least important indicators.

5.1.2 Human Rights

Human rights sub-category consists of seven aspects and 9 indicators. Seven aspects of human rights are investment and procurement practices, non-discrimination, freedom of association and collective bargaining, child labour, forced and compulsory labour, compliance and grievance practices, security practices, and indigenous rights. Child labour (weight = 3.27), forced and compulsory labour (weight = 3.00), and freedom of association and collective bargaining (weight = 2.84) are the often disclosed aspects. Indigenous rights is the less referred aspect (weight = 1.54) (see Fig. 4).

Among the nine human rights indicators, HR6: operations and significant suppliers aspect is identified as having significant risk for incidents of child labour and

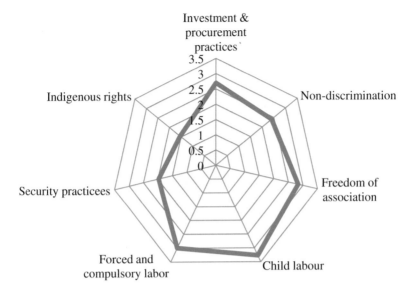

Fig. 4 Aspects of human rights and their scores

measures taken to contribute to the effective abolition of child labour (weight = 3.27); HR7: operations and significant suppliers identified as having significant risk for incidents of forced or compulsory labour and measures taken to contribute to the elimination of all forms of forced or compulsory labour (weight = 3.00); and HR2: percentage of significant suppliers, contractors, and other business partners that have undergone human rights screening, and actions taken (weight = 2.89) are the three most important indicators for apparel firms. The least important human rights indicator is HR9: total number of incidents of violations involving rights of indigenous people and actions taken (weight = 1.54).

5.1.3 Society

The society sub-category is consisted of five aspects and eight indicators. The five aspects of society sub-category are community rights, rights against corruption, public policy, anti-competitive behaviour, and compliance. Community (weight = 2.50), corruption (weight = 2.26), and anti-competitive behaviour (weight = 2.30) are the most disclosed aspects of society. Compliance (weight = 1.59) and public policy (weight = 1.99) are less referred aspects of society sub-category (refer Fig. 5).

The three important indicators of society are SO3: percentage of employees trained in organization's anti-corruption policies and procedures (weight = 2.51); SO1: percentage of operations with implemented local community engagement, impact assessments, and development programs (weight = 2.46); and SO8:

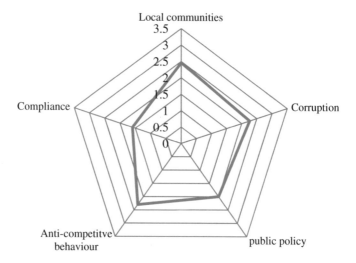

Fig. 5 Aspects of society and their scores

monetary value of significant fines and total number of non-monetary sanctions for non-compliance with laws and regulations (weight = 2.35). SO9: Operations with significant potential or actual negative impacts on local communities (weight = 1.16) are the least important indicators of society sub-category.

5.1.4 Product Responsibility

Product responsibility sub-category consists of five aspects and nine indicators. The five aspects of product responsibility are customer health and safety, product and service labelling, marketing communications, customer privacy, and compliance. Customer health and safety (weight = 2.65) and product and service labelling (weight = 2.40) are the most important aspects, and compliance (weight = 2.19) and market communication (weight = 2.20) are the least referred aspects of product responsibility sub-category (see Fig. 6).

   Product responsibility mean scores ranging between 3.00 and 1.86 with PR1: life cycle stages in which health and safety impacts of products and services are assessed for improvement, and percentage of significant products and services categories subject to such procedures (weight = 3.00) the most referred indicator followed by PR4: total number of incidents of non-compliance with regulations and voluntary codes concerning product and service information and labelling, by type of outcomes (weight = 2.56). PR6: Programs for adherence to laws, standards, and voluntary codes related to marketing communications, including advertising, promotion, and sponsorship (weight = 1.86), are the least important product responsibility indicator.

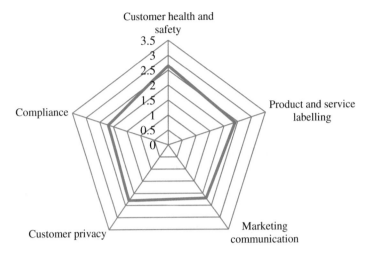

Fig. 6 Aspects of product responsibility and their scores

Overall, social disclosures of apparel firms range between 1.54 and 3.32. Among social sub-categories, labour practices is the more often referred and society is least referred. At aspect level, it is identified that the diversity and equal opportunity of labour and decent work (weight = 3.32) and child labour of human rights (weight = 3.27) are the most often referred. Indigenous rights (weight = 1.54) of human rights and compliance of society (weight = 1.29) are the least important indicators of apparel firms (refer to Fig. 7).

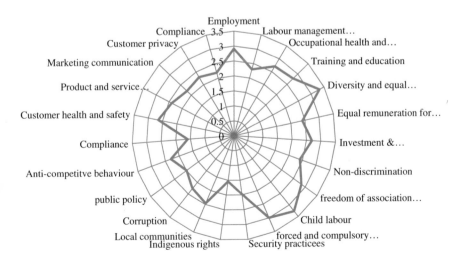

Fig. 7 Factors of social category and their scores

5.2   Level of Social Disclosure and Firm's Characteristics

The following section provides the analysis of the firm's characteristics and their relationship with level of disclosures.

5.2.1   Level of Firm Social Disclosure and Size

Results of the analysis on the level of social disclosure against the firm size are discussed in this section. Firms with the disclosure values of above 'mean plus standard deviation' are considered as the firms with high level of disclosure. Results exhibit that there is only one SME firm and seven of the large firms disclosing social initiatives comprehensively. In terms of relative percentage, SME (20 %) and large (21.8 %) firms are reporting high level of disclosure. Firms with the disclosure values falling between 'mean and mean plus standard deviation' are considered as the firms with the moderate level of disclosure. There is only one SME (20 %) and six large (18.8 %) firms reporting with a moderate level of social disclosure. Finally, value 'below the mean' is considered as low level of social disclosure. Three SMEs (60.0 %) and 19 large (59.4 %) have low level of reporting. Figure 8 shows the distribution of the firms with respect to size and level of disclosure. The results of this analysis exhibit that there is no significant difference among SMEs and large firms with respect to the level of social disclosure.

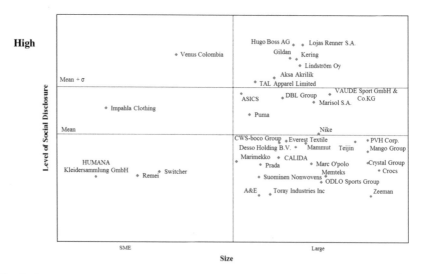

Fig. 8   Firm size and level of social disclosure

Fig. 9 Firm's location and level of social disclosure

5.2.2   Level of Firms Social Disclosure and Location

Results of the analysis on the level of social disclosure against the firm's location are discussed in this section. Of the firms' with moderate level of disclosure (7 firms), there are 2 firms in Asia and European regions and 1 firm in the other three regions: Africa, Latin America, and North America. With respect to the high level of disclosure (8 firms), 3 firms in Europe region, 2 firms in Asia and Latin American regions, and 1 firm belong to North American region. Low level of disclosure (22 firms) is predominantly from Europe 14 firms, followed by 5 firms from Asia and 3 firms from North America. Results demonstrate that the level of disclosure is relatively less in European region and high in Latin American region. Firms from Asian and African regions are moderately disclosing. The mapping of sample firm location against the level of disclosure can be seen in Fig. 9.

6   Conclusions

This paper provides insights on the important social sub-categories of apparel firms in sustainable reports. Results indicate that out of the four sub-categories of social disclosures, labour practices and decent work, and human rights are the most disclosed social sub-categories, whereas the society sub-category is least disclosed among apparel firms. The low scores of society sub-category are due to the lack of importance given by the organization to the public policies in respective countries. Human rights sub-category is another more frequently disclosed sub-category. The importance of human rights sub-category is mainly due to the child labour aspect of human rights. Since the Nike incident in 1991, child labour became critical issues in the apparel industry.

This study also investigates the relationship between firm size and social sustainability reporting. The results indicate that majority of the firms reporting social disclosures are large companies. These results are consistent with earlier studies that the larger firms are likely to publish social reporting. In this study, the proportion of the firms disclosing social indicators is consistent among SMEs and large. Hence, it demonstrates that there is no relationship between firm size and level of disclosure. This study also provides insights on the firm's location and level of disclosure. The results exhibit that the organizations in European region are publishing reports more extensively compared to other regions. The results of this study demonstrate the potential areas that the managers in the apparel manufacturing firms should concentrate to develop socially sustainable firms.

References

Aras, G., Aybars, A., & Kutlu, O. (2010). Managing corporate performance: Investigating the relationship between corporate social responsibility and financial performance in emerging markets. *International Journal of productivity and Performance management, 59*(3), 229–254.

Awaysheh, A., & Klassen, R. D. (2010). The impact of supply chain structure on the use of supplier socially responsible practices. *International Journal of Operations & Production Management, 30*(12), 1246–1268.

Barkemeyer, R., Preuss, L., & Lee, L. (2015). On the effectiveness of private transnational governance regimes—Evaluating corporate sustainability reporting according to the global reporting initiative. *Journal of World Business, 50*(2), 312–325.

Bewley, K., & Li, Y. (2000). Disclosure of environmental information by Canadian manufacturing companies: A voluntary disclosure perspective. *Advances in environmental accounting and management, 1*(1), 201–226.

Bhardwaj, V., & Fairhurst, A. (2010). Fast fashion: Response to changes in the fashion industry. *The International Review of Retail, Distribution and Consumer Research, 20*(1), 165–173.

Chen, L., Feldmann, A., & Tang, O. (2015). The relationship between disclosures of corporate social performance and financial performance: Evidences from GRI reports in manufacturing industry. International Journal of Production Economics.

Cho, C. H., & Patten, D. M. (2007). The role of environmental disclosures as tools of legitimacy: A research note. *Accounting, Organizations and Society, 32*(7), 639–647.

Colicchia, C., Marchet, G., Melacini, M., & Perotti, S. (2013). Building environmental sustainability: Empirical evidence from Logistics Service Providers. *Journal of Cleaner Production, 59*, 197–209.

Ellerup Nielsen, A., & Thomsen, C. (2007). Reporting CSR-what and how to say it? *Corporate Communications: An International Journal, 12*(1), 25–40.

Gereffi, G., & Frederick, S. (2010). The global apparel value chain, trade and the crisis: challenges and opportunities for developing countries. *World Bank Policy Research Working Paper Series.*

GRI (Global reporting Initiative). (2015). Sustainability disclosure database: Data Legend. Available at: https://www.globalreporting.org/SiteCollectionDocuments/GRI-Data-Legend-Sustainability-Disclosure-Database-Profiling.pdf. Accessed on 7 May 2016.

Knebel, S., & Seele, P. (2015). Quo vadis GRI? A (critical) assessment of GRI 3.1 A+ non-financial reports and implications for credibility and standardization. *Corporate Communications: An International Journal, 20*(2), 196–212.

Kozlowski, A., Bardecki, M., & Searcy, C. (2012). Environmental impacts in the fashion industry. *Journal of Corporate Citizenship, 2012*(45), 16–36.

KPMG International Survey of Corporate Responsibility Reporting. (2011). Available at: https://www.kpmg.com/PT/pt/IssuesAndInsights/Documents/corporate-responsibility2011.pdf. Accessed on 7 May 2016.

Lii, Y. S., & Lee, M. (2012). Doing right leads to doing well: When the type of CSR and reputation interact to affect consumer evaluations of the firm. *Journal of Business Ethics, 105* (1), 69–81.

Meng, X. H., Zeng, S. X., Shi, J. J., Qi, G. Y., & Zhang, Z. B. (2014). The relationship between corporate environmental performance and environmental disclosure: An empirical study in China. *Journal of Environmental Management, 145*, 357–367.

Moore, L. L., De Silva, I., & Hartmann, S. (2012). An investigation into the financial return on corporate social responsibility in the apparel industry. *Journal of Corporate Citizenship, 45*, 104–122.

Park-Poaps, H., & Rees, K. (2010). Stakeholder forces of socially responsible supply chain management orientation. *Journal of Business Ethics, 92*(2), 305–322.

Patten, D. M. (1992). Exposure, legitimacy, and social disclosure. *Journal of Accounting and Public Policy, 10*(4), 297–308.

Patten, D. M. (2002). The relation between environmental performance and environmental disclosure: a research note. *Accounting, Organizations and Society, 27*(8), 763–773.

Patten, D. M., & Zhao, N. (2014). Standalone CSR reporting by US retail companies. *Accounting Forum, 38*(2), 132–144.

Pérez, A. (2015). Corporate reputation and CSR reporting to stakeholders: Gaps in the literature and future lines of research. *Corporate Communications: An International Journal, 20*(1), 11–29.

Perry, P., & Towers, N. (2013). Conceptual framework development: CSR implementation in fashion supply chains. *International Journal of Physical Distribution & Logistics Management, 43*(5–6), 478–501.

Qiu, Y., Shaukat, A., & Tharyan, R. (2014). Environmental and social disclosures: Link with corporate financial performance. The British Accounting Review.

Ruwanpura, K. N., & Wrigley, N. (2011). The costs of compliance? Views of Sri Lankan apparel manufacturers in times of global economic crisis. *Journal of Economic Geography, 11*(6), 1031–1049.

Turker, D., & Altuntas, C. (2014). Sustainable supply chain management in the fast fashion industry: An analysis of corporate reports. *European Management Journal, 32*(5), 837–849.

Tschopp, D., & Huefner, R. J. (2015). Comparing the evolution of CSR reporting to that of financial reporting. *Journal of Business Ethics, 127*(3), 565–577.

WBCSD. (2002). Sustainable development reporting: Striking. The Balance World Business Council for Sustainable Development, Geneva, Switzerland

Wensen, K. V., Broer, W., Klein, J. & Knopf, J. (2011). The state of play in sustainability reporting in the EU. Study for the European Commission by CREM BV and adelphi Consultant.

World Trade Organization (WTO). (2012). Available at: http://www.wto.org/english/res_e/statis_e/its2012_e/its12_highlights2_e.pdf. Accessed on 6 May 2016.

World Trade Organization (WTO). (2015). Available at: https://www.wto.org/english/res_e/statis_e/its2015_e/its15_toc_e.htm. Accessed on 6 May 2016.

Zeng, S. X., Xu, X. D., Dong, Z. Y., & Tam, V. W. (2010). Towards corporate environmental information disclosure: an empirical study in China. *Journal of Cleaner Production, 18*(12), 1142–1148.

Printed in the United States
By Bookmasters